Integrating Computational Thinking Through Design-Based Learning

Mustafa Saritepeci · Hatice Yildiz Durak
Editors

Integrating Computational Thinking Through Design-Based Learning

Strategies for Integration in Different Disciplines

Springer

Editors
Mustafa Saritepeci
Department of Instructional Technology
Faculty of Education
Necmettin Erbakan University
Konya, Türkiye

Hatice Yildiz Durak
Department of Instructional Technology
Faculty of Education
Necmettin Erbakan University
Konya, Türkiye

ISBN 978-981-96-0855-3 ISBN 978-981-96-0853-9 (eBook)
https://doi.org/10.1007/978-981-96-0853-9

This Springer imprint is published by the registered company Springer Nature Singapore Pte Ltd.
The registered company address is: 152 Beach Road, #21-01/04 Gateway East, Singapore 189721, Singapore

Introduction

Computational thinking (CT) has become essential for problem-solving and innovation in today's rapidly changing technological landscape. However, ensuring the sustainability of CT instruction is a significant challenge for educators.

This book presents an innovative way of integrating CT from a design-based learning (DBL) perspective. DBL is a transformative approach that combines the power of problem-based learning with the hands-on nature of project-based learning.

CT is inherently about understanding and solving problems, which aligns it with DBL. This book provides detailed guidance on integrating CT concepts into DBL activities. It also explores how AI-supported design-oriented learning activities can be used to develop CT skills.

Importantly, this book provides practical examples of integrating CT into teaching-learning processes in different contexts. These examples range from science, technology, engineering, and mathematics to foreign language education.

Integrating CT from a DBL perspective not only enhances students' ability to problem solve and innovate but also gives them the twenty-first-century skills they need to succeed in the future labor market.

This book will be a valuable resource for researchers, educators, and anyone interested in improving CT teaching. By integrating CT with DBL, we can help our students prepare for future challenges and create innovative solutions.

The book is divided into chapters, each focusing on a different aspect of CT integration through DBL. What follows is a brief overview of each chapter:

Chapter 1, entitled "Introduction to Computational Thinking and Design-Based Learning," authored by Gül Özüdoğru, examines in-depth theoretical knowledge of computational thinking and design-based learning. This chapter discusses the definition, characteristics, scope, related concepts, integration into curricula, and dimensions of computational thinking. It then examines the definition, characteristics, scope, related concepts, processes, and dimensions of design-based learning. Finally, the impact of design-based learning environments on computational thinking skills is discussed. This chapter aims to provide information about computational thinking, design-based learning, and their connection and use in educational environments. This chapter will guide educators and those who will research this subject.

Chapter 2, "Theoretical Framework for Integrating Computational Thinking in Education," authored by Tuğba Kamalı-Arslantaş, addresses effective frameworks for integrating CT into education, aligning with design-based learning. This chapter examines the critical function that CT plays in education, emphasizing its necessity, importance, and the challenges of its integration. The chapter begins by exploring the evolution of CT as a fundamental cognitive tool, not just confined to programming but essential for navigating the complexities of the modern world. It underscores the transformative impact of CT on problem-solving, future workforce preparation, critical thinking, and creativity across various disciplines. The chapter then delves into practical strategies for integrating CT, discussing three specific frameworks: Project-Based Learning (PBL), Digital Storytelling (DST), and the CT Integration Elements proposed by Malyn-Smith et al. (2018). Each framework offers unique perspectives and approaches to fostering CT skills, catering to the diverse needs of learners. Additionally, the chapter addresses the persistent challenges, such as the lack of a universally accepted definition for CT, practical complexities in diverse contexts, and the need for suitable assessment frameworks. Ultimately, this chapter advocates for collaborative efforts from educators, researchers, and policymakers to overcome these challenges and ensure comprehensive and equitable CT integration in education.

In Chap. 3, Nilüfer Atman Uslu discusses integrating DBL to develop CT skills in educational systems. In order to prepare the ground for future generations who can produce digital solutions, it is critical for students to understand the needs of the users of the digital products they will create and to develop solutions for these needs. In this context, DBL provides a framework for teachers, emphasizing five basic principles: authentic learning environments, student-centered approach, inquiry, prototyping, and encouraging deep learning. The advantages and disadvantages of DBL are examined, and its role in educational processes is discussed in more depth. In addition, methodological processes supported by the stages of design thinking are concretized with applicable examples for students to develop CT skills. Finally, Dr. Atman Uslu emphasizes the transformative potential of bringing DBL and CT together in education and draws attention to the opportunities offered at the intersection of these two approaches. This chapter provides valuable information on how educators and students can effectively use these concepts.

Chapter 4, "Integrating Computational Thinking via AI-Based Design-Based Learning Activities," by Mustafa Saritepeci, Hatice Yildiz Durak, and Esra Kidiman explores the integration of CT as an essential skill for solving complex problems across disciplines through a DBL approach. The chapter emphasizes AI's role in enhancing CT skills via AI-supported DBL activities. It proposes a framework based on the TPACK model, incorporating knowledge domains of DBL, content, and AI tools. A detailed seven-step learning activity plan is presented and illustrates the integration of AI-supported DBL into CT instruction. The chapter also discusses AI applications in education, including collaboration tools, personalized feedback, and AI-supported learning frameworks.

Chapter 5, authored by Ali Duran, focuses on integrating CT with computer-assisted language learning (CALL) in foreign language education. It starts with a historical overview of technology's role in language education, emphasizing the transformative impact of computers. This context sets the stage for understanding CALL, which utilizes computer technologies to enhance language learning environments. The chapter critically examines three theoretical perspectives: structural, cognitive/constructivist, and sociocognitive. Each perspective highlights the evolution of language learning theories in conjunction with ICT, moving from drill-based methods to more interactive, communicative approaches. Central to this discussion is DBL, presented as a contemporary application of sociocognitive theory. By incorporating CT-driven DBL activities, the chapter showcases a dynamic and interactive learning experience that equips learners for real-world communication. This approach demystifies the complexity of teaching CT and demonstrates how DBL can foster cognitive engagement through iterative design processes. The chapter concludes with practical guidance for educators, advocating for innovative methodologies that combine CALL and CT. This integration enhances language learning and aligns with the cognitive demands of the modern digital world, offering a forward-thinking approach to education.

Chapter 6, authored by Ali Duran, focuses on the practical application of integrating CT with DBL in foreign language education. The chapter investigates how these innovative pedagogical strategies can enhance language learning outcomes by thoroughly exploring theoretical frameworks and implementation examples. This chapter begins by examining the theoretical underpinnings of CT and DBL, emphasizing their potential to transform traditional language education. Analyzing a selection of case studies from existing literature illustrates the successful application of CT principles within DBL contexts, providing concrete examples of how these methodologies can be employed in classroom settings. The execution phase is meticulously detailed, showcasing how CT processes such as problem decomposition, pattern recognition, and algorithmic thinking can be integrated into language learning tasks. The chapter further discusses the importance of professional development for educators, highlighting the need for training in CT concepts and their application within DBL frameworks. Additionally, the chapter explores various formative and summative assessment techniques to evaluate the effectiveness of CT-driven DBL projects. It emphasizes the role of ongoing evaluation in refining instructional strategies and enhancing learning outcomes. Finally, the chapter concludes with insights into the future directions of CT-driven DBL, advocating for continued research and the development of innovative teaching strategies.

In Chap. 7, Uğur Ferhat Ermiş and Hasan Celal Balıkçı explore the integration of CT in education through DBL, focusing on digital storytelling (DS). The study presents a DS project designed with activities to achieve the objectives. The analysis identifies key themes, including problem decomposition, pattern recognition, abstraction, and algorithm generation. The findings highlight the significant contributions of DS in supporting learning, enhancing educational outcomes, and fostering digital literacy skills. However, the study also identifies several obstacles that impede effective ICT integration, such as limited access to resources, insufficient technical support, and inadequate teacher training. The chapter emphasizes the importance of using DBL with a clear educational purpose, viewing technology as a facilitator. It advocates for a comprehensive understanding of DBL's transformative potential, stressing the need for systematic educational technology integration, effective instructional methods, and active student engagement. In summary, Chap. 7 demonstrates how DS can facilitate the integration of ICT into curricula through DBL, outlining both the advantages and challenges of this approach for educators and students.

Chapter 8, entitled "Computational Thinking Integration by Design-Based Learning Perspective Implementation Example of Math Education," authored by Deniz Kaya and Yıldız Özaydın Aydoğdu, examines the integration of computational thinking skills in mathematics education through a design-based learning (DBL) lens, emphasizing the analysis of application processes and the examination of illustrative examples. This section initially examines the applications utilized in DBL to advance computational thinking abilities within the existing literature. Then, the steps of activity design, determining the outcome(s), target audience characteristics, environmental factors, knowledge and skills that need to be learned, creation of real-life based problem situations, the teacher's roles in learning processes, final problem-solving projects, and evaluation process are given below. The article also addresses how and for what purpose real-life contexts should be utilized, which are essential to DBL processes. Furthermore, the benefits of collaboration in these contexts and the optimal approaches are also delineated. Finally, the CT-driven DBL activity plan is included in the section and can be designed individually or collaboratively. This chapter will guide educators who want to develop computational thinking skills using DBL in mathematics education.

Chapter 9, entitled "Computational Thinking Integration by Design-Based Learning Perspective Implementation Example of AI," authored by Şeyhmus Aydoğdu, presents an illustrative example of the implementation of artificial intelligence in the context of teaching activities designed according to the design-based learning approach to the development of computational thinking skills. This section discusses the importance of developing computational thinking skills among teacher candidates. It then examines research related to teaching artificial intelligence to teacher candidates. Furthermore, it establishes a relationship between the implementation steps of machine learning and computational thinking skills. Finally, it applies a teaching process that incorporates the principles of design-based learning environments to this relationship. A statistically significant increase in the creativity dimension of computational thinking skills was observed during the teaching process. Based on the literature, researchers and practitioners were presented with recommendations in this section. At the end of the section, examples of AI-based activities were provided. This section aims to provide insight into the development of activities designed to enhance computational thinking abilities by integrating artificial intelligence teaching with design-based learning environments.

Chapter 10, "Promoting Computational Thinking Through STEM Education," authored by Kemal Izci, focuses on supporting CT concepts within the STEM fields. This chapter explains the importance of CT and STEM for preparing future citizens. Then, the chapter first discusses how CT and STEM are related and support each other as both concepts involve modeling, reasoning, and problem-solving. Secondly, this chapter provides ways to integrate CT in the STEM context to promote students learning. For the second aim, the chapter mentions three main approaches for integration of CT in K-12 education: (1) introducing programming activities that are separate from content learning, (2) connecting CT with content learning by utilizing problem-solving scenarios to describe, compare, and test predictions about systems, and (3) providing students with insights into how STEM professionals apply CT. Regardless of these approaches, the chapter also explains: (a) instructional strategies, including problem-based learning, experiential learning, and game-based learning; (b) plugged and unplugged instructional tools, including Scratch, Lego, card sorting, and Robotics; and (c) assessments that can be used in STEM context to promote CT skill. Lastly, the chapter discusses the requirements that STEM teachers must possess to integrate CT into STEM instruction successfully.

Chapter 11, written by Effransia Tzagkaraki, Stamatios Papadakis, and Michail Kalogiannakis, examines a study in which physical programming and block-based pedagogical techniques were used to teach electrical circuits to fifth-grade students. Design-based learning provides a framework for evaluating the effects on students' attitudes, motivation, and self-efficacy toward physics. The study observed that these methods, supported by the Micro:bit board, created significant differences in the experimental group students. Moreover, the positive impact of the intervention on student attitudes supports the endeavor to improve the pedagogical approach in physics and ultimately to strengthen computational thinking skills.

This book ends with Chap. 12, "Assessment of Computational Thinking Skills" by Hüseyin Can ŞENEL and Selma ŞENEL, which presents and discusses the various approaches developed to assess computational thinking skills. The initial approach will concentrate on scales based on the individual's perception of their computational thinking abilities. An alternative approach is the utilization of maximum performance tests for the assessment of computational thinking abilities. Performance-based assessment and blended assessment systems are other methods covered in this chapter. This chapter will introduce the approaches mentioned above and present examples from the literature. Furthermore, common and eligible examples will be discussed in the context of design-based learning. This chapter will guide readers with a good review of the best examples, an example rubric developed by the authors, and advice for measurement methods for a complex structure like CT.

Throughout its chapters, the book presents an integration of theoretical frameworks and practical strategies for integrating IT in various educational contexts. The book examines the role of AI-supported DBL activities in developing IT skills. Moreover, it provides practical examples of ICT integration in science, technology, engineering, mathematics (STEM), and foreign language education. It discusses the potential of digital storytelling as a tool for promoting ICT. The chapters offer detailed guidance on designing and implementing ICT-focused activities and emphasize the transformative potential of combining ICT with DBL in equipping students with the twenty-first-century skills needed for future success. It also emphasizes the importance of ongoing assessment and professional development to improve teaching strategies and learning outcomes in this subject.

This book is valuable for educators, educational leaders, researchers, and stakeholders. By providing research-based, evidence-based recommendations for the use of digital technologies and unplugged activities, it aims to help educators effectively integrate ICT into their teaching practice and prepare students for the challenges and opportunities of the future.

Konya, Türkiye

Mustafa Saritepeci
Hatice Yildiz Durak

Contents

Chapter 1
Introduction to Computational Thinking and Design-Based Learning

Gül Özüdoğru

Abstract With the rapid development of technology, learning environments and the features expected from learners are changing. In the digital age, where constructivist learning systems are based, learners are expected to be active in learning environments. Active participation of students in the learning process also improves their thinking skills. Computational thinking (CT), one of these skills, focuses on utilizing digital technologies in problem-solving processes. Design-based learning (DBL) environments are very important in developing this skill. Design-based learning environments also focus on solving real-world problems and mostly use technology. Knowing the characteristics of DBL environments and planning and applying the processes accordingly will help to achieve successful results. DBL is preferred when centering the learners in the learning processes. When the literature is examined, CT and DBL are discussed in an interdisciplinary context in many fields, such as language, science, mathematics, and technology. Methods such as robotic coding, programming, gamification, and digital storytelling, especially suitable for DBL integrated into these areas, are essential in developing CT skills. These methods ensure that students build their learning processes by doing and experiencing. This part of the book will discuss the definition, scope, characteristics, related concepts, and dimensions of CT. Then, the definition, scope, characteristics, related concepts, and dimensions of DBL will be discussed. Finally, the effect of DBL environments on CT skills will be discussed with the support of the literature. Thus, it aims to reveal the advantages, limitations, difficulties, and connections between CT and DBL in education. This section will likely guide educators and those researching CT skills in DBL environments.

Keywords Computational thinking · Design-based learning · Teaching CT via DBL

G. Özüdoğru (✉)
Department of Educational Science, Kırşehir Ahi Evran University, Kırşehir, Türkiye
e-mail: gulerturk87@gmail.com

M. Saritepeci and H. Yildiz Durak (eds.), *Integrating Computational Thinking Through Design-Based Learning*, https://doi.org/10.1007/978-981-96-0853-9_1

Computational Thinking

Definition of Computational Thinking

Technology has firmly entrenched itself in every facet of life during the digital age. When employed judiciously and efficiently, it offers numerous benefits and conveniences to individuals. Additionally, technology serves as a valuable tool for problem-solving and overcoming challenges. At this juncture, computational thinking (CT) refers to an approach focused on solving problems in a manner suitable for computer implementation (Barr & Stephenson, 2011; Yildiz Durak et al., 2023b). Computers act as agents that execute computational operations, adhering to instruction programs to carry out arithmetic or logic operations. (Denning & Tedre, 2019). Comprehensive knowledge of the terminology employed in computer science is crucial for grasping problems and facilitating the creation of efficient solutions (Kazimoglu et al., 2012). As an academic discipline, computer science possesses a distinct body of knowledge that empowers students to become evaluators, independent learners, and designers of new technologies (Csizmadia et al., 2015).

The seeds of CT have germinated in many directions over the centuries. Still, its development as a professional subject for computer designers and programmers gained momentum with the introduction of electronic computers into the industry in the 1950s (Denning & Tedre, 2019). The central concept driving the CT movement is that the knowledge and skills acquired through computer science have broad applications that can benefit various fields (Ch'ng et al., 2019). CT equips students with knowledge, skills, and a distinctive problem-solving mindset (Csizmadia et al., 2015).

CT is considered to be an essential skill for the twenty-first century. The earliest reference to CT dates back to Papert (1980) in his book on learning to think from childhood onwards, asserting that the vision of integrating CT into everyday life was yet to be fully developed. Still, numerous experiments were on the way, and eventually, it would be widespread. Wing (2006) emphasized that CT, deemed a fundamental skill for computer scientists and everyone, involves problem-solving, system design, and understanding human behavior through the essential concepts of computer science.

CT is defined in various ways in the literature. One perspective, as proposed by Wing (2006), frames it as the reformulation of a challenging problem into one that can be solved using techniques like reduction, embedding, transformation, or simulation. Another viewpoint emphasizes the necessity of understanding how to communicate with computers to harness their computing power effectively, terming successful communication in this context as CT (Denning and Tedre 2019). Shute et al. (2017) assert that CT is a learned skill, emphasizing that our brains do not inherently think computationally. They define CT as the mental skills and practices required for crafting computations that empower computers to execute tasks, thus elucidating and interpreting the world as a complex array of information processes. According to Grover and Pea (2013), the core of CT lies in approaching problems with a mindset akin to that of a computer scientist. According to Csizmadia et al. (2015), CT is a cognitive process characterized by logical reasoning with solving problems and comprehensively understanding artifacts, procedures, and systems.

According to Wing (2006), computer science is the exploration of computation, encompassing what can be computed and the methodologies for computation. Consequently, CT is characterized by the following aspects:

- CT is not merely programming, it involves conceptualization; computer science transcends mere computer programming. Thinking like a computer scientist entails more than simply coding. It necessitates thinking at diverse levels of abstraction.
- CT is not merely a skill based on memorization; it is a fundamental skill in itself. A fundamental skill is something that every human being needs to know to function in modern society.
- CT is the way people think, not the computers. CT is a way for people to solve problems; it does not try to make people think like computers. Computers are boring and dull; people are intelligent and creative. It's people who infuse excitement into computers.
- CT complements and merges mathematical and engineering perspectives. Similar to other sciences, computer science is deeply rooted in mathematics due to its formal underpinnings. Furthermore, computer science inherently draws from engineering principles as it involves creating systems that interact with the physical world. The limitations imposed by the basic computing device compel computer scientists to think in terms of computation, extending beyond pure mathematics.
- CT transcends mere artifacts; it encompasses ideas. Its influence extends beyond the tangible presence of software and hardware artifacts that pervade our lives. It encompasses the computational concepts we utilize to navigate and resolve challenges, organize our daily routines, and engage in communication and collaboration with others.
- CT is for everyone, everywhere. CT will achieve its actual realization when it seamlessly integrates into human endeavors, ceasing to be an open philosophy and becoming an inherent part of our collective approach (Wing, 2006).

Programming tools and environments are used to engage students in CT, an approach to problem-solving that uses the principles of computer science (Yadav et al., 2017; Yildiz Durak, 2020). While CT is more than algorithmic thinking, using CT skills to solve a problem will ultimately involve algorithmic and accompanying programming skills to produce the desired solution (Angeli, 2022). Programming is a fundamental skill of computer science and an important tool to support the cognitive tasks involved in CT, but it is also an indicator of computational proficiencies (Grover & Pea, 2013; Yildiz Durak, 2024). There is a growing interest in integrating CT into K-12 education, driven by its role in fostering thinking skills and digital competencies among children. CT and coding have become an essential component of school curricula in many countries (Angeli & Giannakos, 2020). As a result of the literature review in their research, Bocconi et al. (2016) found that the justifications for teaching CT in K-12 education have two trends: the general benefits of IT as a thinking skill and the need to develop new skills for the employment market. However, the implementation of CT in case studies is affected by the type of CT integration approach, the time dedicated to the development of CT skills, and the collaboration between teachers of the courses in which CT skills are developed (Bocconi et al., 2022). Its applications extend to screen-based computer programming, controlling digital circuits and robots, and, more broadly, designing algorithmic solutions for problems executable by computers (Gadanidis et al., 2017). Keith et al. (2019) emphasized the significance of robotics in advancing CT. CT is not merely a set of concepts for programming but a very rich skill set, involving ways of thinking and practicing that are honed and sharpened through practice (Denning & Tedre, 2019; Durak et al., 2020). Recognizing it as a vital body of knowledge, Angeli and Giannakos (2020) assert that CT is indispensable for students to effectively confront the challenges of the twenty-first century. Incorporating programming activities into curricula is anticipated to enhance students' overall thinking skills.

CT connects various disciplines, with studies exploring its integration into fields such as mathematics and science (Bar & Stephenson, 2011; Gadanidis et al., 2017; Mehalik et al., 2008; Tan et al., 2019). Some studies address CT within the framework of STEM (Science, Technology, Engineering, Mathematics) (Yildiz Durak et al., 2023a; Zhang & Biswas, 2019) and TPACK (Technological Pedagogical Content Knowledge) (Angeli et al., 2016; Kong & Lai, 2022; Saritepeci, 2022; Yildiz Durak et al., 2023a). In addition, CT has an important place in teacher education (Angeli, 2022; Cabrera, 2019; Gadanidis et al., 2017; Kong & Lai, 2022; Yadav et al., 2017). Kong and Lai (2022) underscored the significance of teachers' skills, introducing a CT teacher development model. Cabrera (2019) highlighted the imperative for teachers to incorporate CT into K-12 education. Through a literature review, the study revealed that teachers may harbor biases and preconceptions about learning and teaching CT. One approach to address these biases and preconceptions is to conduct studies to enhance the development of teachers' CT skills. In a bibliographic data analysis study, Tekdal (2021) concluded that research on CT has experienced exponential growth since 2013. This growth is attributed to national and international collaborations among researchers from various sectors, with publications primarily appearing in journals specializing in educational technology. The knowledge base is from education, computer science, and social sciences. Tekdal's findings also revealed three main themes within the research topics: integrating CT into STEM education, experimental studies evaluating CT skills, and discussions on defining CT skills.

Dimensions of Computational Thinking

The literature encompasses studies that investigate the elements, components, or dimensions of CT in diverse ways, providing valuable insights into the framework and scope of CT. The research conducted by Angeli et al. (2016) indicates that at K-6 levels, CT involves a thought process utilizing elements such as abstraction, generalization, decomposition, algorithms, and debugging. Grounded in these five CT skills, the researchers developed a curriculum framework specifying competency indicators for these skills, ranging from simple to complex. The curriculum framework Angeli et al. (2016) developed is designed to prompt children to engage in critical thinking and problem-solving. It encourages them to develop solutions for problems, automate them through algorithms, and generalize the problem-solving approach to new problems by identifying or recognizing common patterns. It also aims to introduce the thinking processes of CT to young students so that they become competent in learning more advanced theoretical and practical computer science topics in secondary education.

The International Society for Technology in Education and the Computer Science Teachers Association (ISTE & CSTA, 2011) define CT as a problem-solving process that includes (but not limited to) the following characteristics:

- The problems are formulated in a manner conducive to their solution through using computers and other tools.
- The data are organized and analyzed logically.
- The data are represented through abstractions such as models and simulations.
- The solutions are automated through algorithmic thinking (a series of sequential steps).
- The possible solutions are identified, analyzed, and implemented to achieve the most efficient and effective combination of steps and resources.
- The problem-solving process is generalized and transferred to a wide variety of problems.

It is also expressed that these skills are supported and developed by a set of dispositions or attitudes that are the basic dimensions of CT. These include confidence in dealing with complexity, tolerance for ambiguity, persistence under challenging problems, ability to deal with open-ended problems, and the ability to communicate and work with others to achieve a common goal or solution (ISTE & CSTA, 2011).

ISTE (2017) stated CT competencies to help students benefit from the power of computing to innovate and solve problems. They aim to guide educators in integrating CT across disciplines with all students. ISTE (2017) made suggestions for learners, leaders, collaborators, designer and facilitators (educators) to improve CT competencies. For students it is as follows: (1) Set professional learning goals (2) Learn how and when to use computation (3) Leverage CT experts and resources (4) Develop a growth mindset (5) Recognize Inequities and Opportunities of CT. For educators, it is as follows: (1) Account for learner variability; (2) Empower student choice; (3) Vary instructional approaches; and (4) Use alternative assessments.

Researchers have approached the dimensions of CT in various ways, and the preference for a specific classification often depends on the intended context for research and applications. The classification of the dimensions of CT in some studies can be expressed as in Table 1.1.

When Table 1.1 is examined, it is striking that there are common categories. Some basic concepts can be explained as Table 1.2 (Angeli et al., 2016; Barr & Stephenson, 2011; Csizmadia et al., 2015; ISTE & CSTA, 2011; Lee et al., 2014; Kong, 2019).

Design-Based Learning

Definition of Design-Based Learning

Design-based learning holds a significant role in contemporary constructivist learning systems. This educational approach centers on inquiry and reasoning processes to generate innovative artifacts, systems, and solutions (Gomez Puente et al., 2013a). Integrating design thinking and processes, DBL is applicable across various subjects in the curriculum (Hsu et al., 2018). This methodology engages students in intricate, often interconnected research and design processes characterized by student-centered, design-focused, and inquiry-oriented dynamic practices (Zhang et al., 2022). Such learning environments contribute to integrating twentyfirst-century skills into the curriculum, including hands-on work, collaborative teamwork, problem-solving, active learning, creative and innovative designs, and participation in real-world tasks (Chandrasekaran et al., 2013).

DBL has advantages such as providing students with a rich understanding of design and technology, an active learning environment, collaborative learning opportunities, and motivation (Doppelt et al., 2008). Similarly, Weng et al. (2023) concluded that design-based learning activities improve students' motivation, active participation, and higher-order thinking skills, such as problem-solving and critical thinking, and encourage deep learning. However, due to the open-ended nature of design, navigating the content, the design process, and the need for teamwork skills may create too much cognitive load for low-achieving students and leave them behind (Doppelt et al., 2008). The system design-based approach allows students to express their own needs, bringing them closer to asking questions worth investigating (Mehalik et al., 2008). DBL helps students understand the complexity of the actual design process as they deal with real-world problems (Doppelt, 2009).

DBL is derived from project-based learning and problem-based approaches by adopting student-centered educational principles (Gomez et al., 2011a, 2011b). Students can use project-based learning to solve design problems, and students with different learning styles can express their skills and abilities by working on projects in authentic learning environments (Chandrasekaran et al., 2013). A design process is similar to problem-solving with its general six-stage structure: defining the problem and identifying the need; gathering information; presenting alternative solutions; selecting the most appropriate solution; designing and building a prototype; and evaluating and refining the process (Doppelt and Schunn, 2008). DBL encourages students to establish their objectives and generate ideas to accomplish these goals (Hsu et al., 2018). By utilizing a natural and meaningful environment, DBL not

Table 1.1 Dimensions of CT by some research

ISTE and CSTA (2011)	Bar and Stephenson (2011)	Berland and Lee (2011)	Kazimoglu et al. (2012)	Lee et al. (2014)	Csizmadia et al. (2015)	Angeli et al. (2016)	Kong (2019)
Data collection	Data collection	Conditional logic	Problem-solving	Algorithmic thinking	Algorithmically	Abstraction	Abstraction/ abstracting, modelling/ abstracting and modularising
Data analysis	Data analysis	Algorithm building	Building algorithms	Decomposition	Decomposition	Generalization	Algorithmic thinking
Data presentation	Data presentation	Debugging	Debugging	Pattern recognition	Generalization	Decomposition	Testing and debugging
Problem decomposition	Problem decomposition	Simulation	Simulation	Pattern generalization and abstraction	Pattern	Algorithms	Being incremental and iterative
Abstraction	Abstraction	Distributed computation	Socialising	Unarticulated instances	Abstraction	Debuging	Problem decomposition
Algorithms and procedures	Algorithms and procedures				Presentation		Planning and designing
Automation	Automatization				Evaluation		Reusing and remixing
Simulation	Parallelization						
Parallelization	Simulation						

Table 1.2 Explanation of CT dimensions

Dimension	Explanation
Data collection	The process of collecting convenient information
Data analysis	The process of making sense of the data, finding patterns and drawing conclusions
Data presentation	The process of describing and organizing data with appropriate charts, words, graphs or pictures
Problem decomposition	The process of breaking down a large problem into smaller sub-problems or details. Objects and methods are defined, the main and functions are defined. Decomposing problems into subtasks is a high-level abstraction process; it is breaking down tasks into manageable, smaller parts
Abstraction	Using procedures to contain a series of frequently repeated commands that fulfill a function, reducing complexity to identify the main idea
Algorithms (Algorithmic thinking)	The logical steps required to create a solution to a specific problem. A sequential series of steps taken to solve a problem or achieve a result
Debugging	Recognizing when instructions do not correspond to actions and correcting errors. In short, detecting and correcting errors
Automation	The use of machines or computers to perform tedious or repetitive tasks. Use of various computer and technology tools
Generalization	The solution to similar types of problems arises from past experience in solving such problems
Simulation	A model or representation of a process. Simulation also involves conducting experiments using models
Evaluation	The process of making sure that a solution is good, i.e. fit for purpose
Parallelization	Resources are organized to perform tasks simultaneously to achieve a common goal

only enhances students' understanding of the content but also fosters the acquisition of design skills, nurtures creativity, and initiates the learning process aligned with individual preferences, learning styles, and diverse skills (Doppelt et al., 2008).

Dimensions of Design-Based Learning

Design thinking is important in DBL. Brown (2008) emphasized the importance of creativity, bright minds, and great ideas in design thinking and defined the design process metaphorically as a system of fields rather than a series of predefined, orderly steps. Brown (2008) also stated that design projects should go through three areas: Inspiration, Ideation, and Implementation. According to Prommuna and Kantathanawatb (2020), in DBL, learning activities have an interdisciplinary integration approach that requires creative, interactive, technological innovation to systematically solve problems. They classified the DBL process into six stages in his research: Identify needs, design review, collect information, prototype, production, and evaluate.

Zhang et al. (2022) analyzed DBL activities, considering four interconnected dimensions. The first dimension is the learning activity, focusing on the sequence and ritual of the design and learning process, task execution, and student learning approaches. The second dimension is the grouping strategy, which addresses allocating learning tasks and whether students work individually or in small groups. The third dimension is social interaction, encompassing peer-to-peer communication and student–teacher interaction. The fourth dimension is the temporal aspect, relating to the duration and sequence of learning tasks (Zhang et al., 2022).

Drawing from the literature, Gomez Puente et al. (2013b) framed the characteristics of DBL across five dimensions: project characteristics, design elements, teacher's role, assessment, and social context. They explained these as follows: Authentic and open-ended scenarios, hands-on experiences and integrating multidisciplinary in the dimension of project characteristics. Exploring problem representation and using interactive/iterative design methodology in the dimension of design elements. Coaching on task, process and personal development as a teacher's role. Formative and summative (process and product) evaluation in the dimension of assessment. Peer-to-peer activities, communication and competitions for collaborative learning in the dimension of social context (Gomez Puente et al., 2013b).

The Effect of Design-Based Learning Environments on Computational Thinking

In educational settings, lesson planning aligns with real-life problems. Designing tasks for real-life problem-solving can be challenging, prompting curriculum designers to employ rapid prototyping before developing an entire curriculum, course, or module (Angeli et al., 2016). Within CT, each learning task is crafted to be authentic, pertinent to children's lives, and introduces learners to genuine CT experiences (Angeli et al., 2016). As Csizmadia et al. (2015) emphasized, CT involves cultivating thinking skills and provides essential support for learning and comprehension. DBL activities such as digital storytelling, interactive web page design, and the creation of infographic design can have essential results in teaching CT and reflecting it (Saritepeci, 2020). DBL in computational literacy is effective for creative CT, in which students create their ideas and make meaningful works from them (Kim & Han, 2012).

Since programming and algorithm courses tend to be challenging and tedious in developing CT, it is crucial to apply DBL based on constructivist theory in these courses (Jun et al., 2017). DBL activities supporting the efficient use of digital technologies can significantly affect CT skills (Saritepeci, 2020). Design-based learning is a self-directed approach where students partake in self-guided learning by crafting creative and innovative solutions that align with academic, community, and industry expectations (Chandrasekaran et al., 2013). Jun et al. (2017) categorized the steps of DBL in the context of CT into four key headings design, which involves the process of creating and interacting rather than simply using; personalization, encompassing the creation of meaningful and personally relevant content; collaboration, which involves the creation of materials in collaboration with others and the generation of ideas about the material and reflection, the process of reviewing, revising, and evaluating materials.

Numerous studies have explored the impact of DBL environments on CT. In their investigation into the influence of game programming on CT skill development, Brennan and Resnick (2012) suggested that DBL activities can enhance CT skills. In an experimental study on programming instruction, Jun et al. (2017) concluded that DBL led to improvements in the CT skills of primary school students. Similarly, Saritepeci (2020) investigated the effect of DBL activities and programming education on CT skill levels with an experimental study. His research observed a significant positive development in CT skills in both groups. However, Matere et al. (2023) found in an experimental study comparing design-based and traditional learning that there was no significant difference in CT skills between the two groups. Additionally, students in the DBL approach achieved significantly higher skills in final product design than those in the traditional approach. However, a systematic literature review conducted by Hsu et al. (2018), which examined studies on the learning and teaching of CT skills between 2006 and 2018, concluded that DBL activities were among the less preferred approaches for developing CT skills. Despite varied findings, the overall consensus suggests that the effectiveness of DBL in fostering CT may depend on specific contexts and instructional methodologies. Prommuna and Kantathanawatb (2020) developed an integrated design-based learning model based on the computational thinking of undergraduate students with participatory action research.

DBL emphasizes student participation in developing, creating, and evaluating the designed work (Prommuna & Kantathanawatb, 2020). Denning & Tedre, 2019, stated that one of the basic concepts of CT is design and that CT is no longer limited to just programs and algorithms in problem-solving but also focuses largely on design. The use of DBL environments may be beneficial in developing students' computational thinking skills.

Conclusion

This section includes in-depth theoretical information about CT and DBL. CT is one of the basic skills expected from individuals in the twenty-first century. There are different definitions of CT in the literature. In summary, it can be said that CT is an approach that makes use of computer science to solve problems. The dimensions of CT are generally addressed as data collection, data analysis, data presentation, problem decomposition, abstraction, algorithms, debugging, automation, generalization, simulation, evaluation, and parallelization. It is seen that CT teaching is being integrated into curricula, especially in terms of the development of students' thinking skills and technological skills. However, it is important for educators to know how to cooperate and implement it. DBL, on the other hand, is an approach that supports twentyfirst-century skills, is in line with the constructivist approach, and makes students active. It has been stated that the dimensions of DBL are the characteristics of the project, design elements, the role of the teacher, assessment, and social context. It can be stated that DBL environments are supportive of the development of CT.

References

Angeli, C. (2022). The effects of scaffolded programming scripts on pre-service teachers' computational thinking: Developing algorithmic thinking through programming robots. *International Journal of Child-Computer Interaction, 31*, 100329. https://doi.org/10.1016/j.ijcci.2021.100329

Angeli, C., & Giannakos, M. (2020). Computational thinking education: Issues and challenges. *Computers in Human Behavior, 105*, 106185. https://doi.org/10.1016/j.chb.2019.106185

Angeli, C., Voogt, J., Fluck, A., Webb, M., Cox, M., Malyn-Smith, J., & Zagami, J. (2016). A K-6 computational thinking curriculum framework: Implications for teacher knowledge. *Journal of Educational Technology & Society, 19*(3), 47–57.

Barr, V., & Stephenson, C. (2011). Bringing computational thinking to K-12: What is Involved and what is the role of the computer science education community? *ACM Inroads, 2*(1), 48–54.

Berland, M., & Lee, V. R. (2011). Collaborative strategic board games as a site for distributed computational thinking. *International Journal of Game-Based Learning (IJGBL), 1*(2), 65–81. https://doi.org/10.4018/ijgbl.2011040105

Bocconi, S., Chioccariello, A., Dettori, G., Ferrari, A., Engelhardt, K., Kampylis, P., & Punie, Y. (2016, June). Developing computational thinking: Approaches and orientations in K-12 education. In *EdMedia+ innovate learning* (pp. 13–18). Association for the Advancement of Computing in Education (AACE).

Bocconi, S., Chioccariello, A., Kampylis, P., Dagienė, V., Wastiau, P., Engelhardt, K., Earp, J., Horvath, M., Jasutė, E., Malagoli, C. and Masiulionytė-Dagienė, V., & Stupurienė, G. (2022). In A. Inamorato dos Santos, R. Cachia, N. Giannoutsou, & Y. Punie (Eds.), *Reviewing computational thinking in compulsory education*. Publications Office of the European Union. https://doi.org/10.2760/126955

Brown, T. (2008). Design thinking. *Harvard Business Review*, 1–10.

Brennan, K., & Resnick, M. (2012). New frameworks for studying and assessing the development of computational thinking. In *Proceedings of the Annual Meeting of the American Educational Research Association,* Canada, pp. 1–25.

Cabrera, L. (2019). Teacher preconceptions of computational thinking: A systematic literature review. *Journal of Technology and Teacher Education, 27*(3), 305–333.

Ch'ng, S. I., Low, Y. C., Lee, Y. L., Chia, W. C., & Yeong, L. S. (2019). Video games: A potential vehicle for teaching computational thinking. In S.-C. Kong, & H. Abelson (Eds.), *Computational thinking education* (pp. 247–260). Springer Open.

Chandrasekaran, S., Stojcevski, A., Littlefair, G., & Joordens, M. (2013). Project-oriented design-based learning: Aligning students' views with industry needs. *International Journal of Engineering Education, 29*, 1109–1118.

Csizmadia, A., Curzon, P., Dorling, M., Humphreys, S., Ng, T., Selby, C., & Woollard, J. (2015). *Computational thinking-A guide for teachers.* Computing at School.

Denning, P. J., & Tedre, M. (2019). *Computational thinking.* MIT Press Essential Knowledge Series.

Doppelt, Y. (2009). Assessing creative thinking in design-based learning. *International Journal of Technology and Design Education, 19*(1), 55–65.
Doppelt, Y., Mehalik, M. M., Schunn, C. D., Silk, E., & Krysinski, D. (2008). Engagement and achievements: A case study of design-based learning in a science context. *Journal of Technology Education, 19*(2), 21–28.
Doppelt, Y., & Schunn, C. D. (2008). Identifying students' perceptions of the important of a design-based learning environment. *Learning Environments Research, 11*(3), 195–209.
Durak, H. Y., Saritepeci, M., Topçu, A., & Durak, A. (2020). Investigation of variables related to computational thinking self-efficacy level in middle school students: Are demographic variables, academic success, or programming-related variables more important? In M. Kalogiannakis & S. Papadakis (Eds.), *Handbook of research on tools for teaching computational thinking in P-12 education* (pp. 54–74). IGI Global.
Gadanidis, G., Cendros, R., Floyd, L., & Namukasa, I. (2017). Computational thinking in mathematics teacher education. *Contemporary Issues in Technology and Teacher Education, 17*(4), 458–477.
Gomez Puente, S. M., Van Eijck, M., & Jochems, W. (2011a). Towards characterizing design-based learning in engineering education: A review of the literature. *European Journal of Engineering Education, 36*(2), 137–149.
Gomez Puente, S. M., van Eijck, M., & Jochems, W. (2011b). Towards characterizing design-based learning in engineering education: A review of the literature. *European Journal of Engineering Education, 36*(2), 137–149. https://doi.org/10.1080/03043797.2011.565116
Gomez Puente, S. M., Van Eijck, M., & Jochems, W. (2013a). A sampled literature review of design-based learning approaches: A search for key characteristics. *International Journal of Technology and Design Education, 23*(3), 717–732. https://doi.org/10.1007/s10798-012-9212-x
Gomez Puente, S. M., Van Eijck, M., & Jochems, W. (2013b). Empirical validation of characteristics of design-based learning in higher education. *International Journal of Engineering Education, 29*(2), 491–503.
Grover, S., & Pea, R. (2013). Computational thinking in K–12: A review of the state of the field. *Educational Researcher, 42*(1), 38–43.
Hsu, T. C., Chang, S. C., & Hung, Y. T. (2018). How to learn and how to teach computational thinking: Suggestions based on a review of the literature. *Computers & Education, 126*, 296–310. https://doi.org/10.1016/j.compedu.2018.07.004
ISTE (2017). *Computational thinking competencies.* https://iste.org/standards/computational-thinking-competencies at 06.02.2024 tarihinde erişildi.
ISTE, & CSTA. (2011). *Computational thinking in K–12 education leadership toolkit.* Retrieved February 13, 2024 from https://cdn.iste.org/www-root/2020-10/ISTE_CT_Leadership_Toolkit_booklet.pdf
Jun, S., Han, S., & Kim, S. (2017). Effect of design-based learning on improving computational thinking. *Behavior & Information Technology, 36*(1), 43–53. https://doi.org/10.1080/0144929X.2016.1188415
Kazimoglu, C., Kiernan, M., Bacon, L., & MacKinnon, L. (2012). Learning programming at the computational thinking level via digital game-play. *Procedia Computer Science, 9*, 522–531. https://doi.org/10.1016/j.procs.2012.04.056
Keith, P. K., Sullivan, F. R., & Pham, D. (2019). Roles, collaboration, and the development of computational thinking in a robotics learning environment. In S.-C. Kong, & H. Abelson (Eds.), *Computational thinking education* (pp. 223–245). Springer Open.
Kim, S., & Han, S. (2012). Design-based learning for computational thinking. *Journal of the Korean Association of Information Education, 16*(3), 319–326.
Kong, S. C. (2019). Components and methods of evaluating computational thinking for fostering creative problem-solvers in senior primary school education. In S.-C. Kong, & H. Abelson (Eds.), *Computational thinking education* (pp. 119–141).

Kong, S. C., & Lai, M. (2022). A proposed computational thinking teacher development framework for K-12 guided by the TPACK model. *Journal of Computers in Education, 9*(3), 379–402. https://doi.org/10.1007/s40692-021-00207-7

Lee, T. Y., Mauriello, M. L., Ahn, J., & Bederson, B. B. (2014). CTArcade: Computational thinking with games in school age children. *International Journal of Child-Computer Interaction, 2*(1), 26–33.

Mehalik, M. M., Doppelt, Y., & Schunn, C. D. (2008). Middle-school science through design-based learning versus scripted inquiry: Better overall science concept learning and equity gap reduction. *Journal of Engineering Education, 97*(1), 71–85.

Matere, I. M., Weng, C., Astatke, M., Hsia, C. H., & Fan, C. G. (2023). Effect of design-based learning on elementary students computational thinking skills in visual programming maker course. *Interactive Learning Environments, 31*(6), 3633–3646. https://doi.org/10.1080/10494820.2021.1938612

Papert, S. (1980). *Mindstorms: Children, computers, and powerful ideas*. Basic Books.

Prommuna, P., & Kantathanawatb, T. (2020). The development of an integrated design-based learning model, based on computational thinking for undergraduate students with participatory action research. *Development, 13*(1), 1505–1518.

Saritepeci, M. (2020). Developing computational thinking skills of high school students: Design-based learning activities and programming tasks. *The Asia-Pacific Education Researcher, 29*(1), 35–54. https://doi.org/10.1007/s40299-019-00480-2

Saritepeci, M. (2022). Modelling the effect of TPACK and computational thinking on classroom management in technology enriched courses. *Technology, Knowledge and Learning, 27*(4), 1155–1169. https://doi.org/10.1007/s10758-021-09529-y

Shute, V. J., Sun, C., & Asbell-Clarke, J. (2017). Demystifying computational thinking. *Educational Research Review, 22*, 142–158. https://doi.org/10.1016/j.edurev.2017.09.003

Tan, C. W., Yu, P. D., & Lin, L. (2019). Teaching computational thinking using mathematics gamification in computer science game tournaments. In S.-C. Kong, & H. Abelson (Eds.), *Computational thinking education* (pp. 167–181). Springer Open.

Tekdal, M. (2021). Trends and development in research on computational thinking. *Education and Information Technologies, 26*(5), 6499–6529. https://doi.org/10.1007/s10639-021-10617-w

Weng, C., Chen, C., & Ai, X. (2023). A pedagogical study on promoting students' deep learning through design-based learning. *International Journal of Technology and Design Education, 33*(4), 1653–1674. https://doi.org/10.1007/s10798-022-09789-4

Wing, J. M. (2006). Computational thinking. *Communications of the ACM, 49*(3), 33–35.

Yadav, A., Stephenson, C., & Hong, H. (2017). Computational thinking for teacher education. *Communications of the ACM, 60*(4), 55–62. https://doi.org/10.1007/978-3-319-52691-1_13

Yildiz Durak, H. (2020). The effects of using different tools in programming teaching of secondary school students on engagement, computational thinking and reflective thinking skills for problem solving. *Technology, Knowledge and Learning, 25*(1), 179–195. https://doi.org/10.1007/s10758-018-9391-y

Yildiz Durak, H. (2024). The model for middle school students' computational identity. *Education and Information Technologies, 29*, 4223–4251.

Yildiz Durak, H., Atman Uslu, N., Canbazoğlu Bilici, S., & Güler, B. (2023a). Examining the predictors of TPACK for integrated STEM: Science teaching self-efficacy, computational thinking, and design thinking. *Education and Information Technologies., 28*, 7927–7954. https://doi.org/10.1007/s10639-022-11505-7

Yildiz Durak, H., Saritepeci, M., & Durak, A. (2023b). Modeling of relationship of personal and affective variables with computational thinking and programming. *Technology, Knowledge and Learning, 28*, 165–184. https://doi.org/10.1007/s10758-021-09565-8

Zhang, F., Markopoulos, P., Bekker, T., Paule-Ruíz, M., & Schüll, M. (2022). Understanding design-based learning context and the associated emotional experience. *International Journal of Technology and Design Education, 32*(2), 845–882. https://doi.org/10.1007/s10798-020-09630-w

Zhang, N., & Biswas, G. (2019). Defining and assessing students' computational thinking in learning by modeling environment. In S.-C. Kong, & H. Abelson (Eds.), *Computational thinking education* (pp. 203–221). Springer Open.

Chapter 2
Theoretical Framework for Integrating Computational Thinking in Education

Tuğba Kamalı Arslantaş

Abstract Computational Thinking (CT) encompasses the cognitive processes utilized to articulate problems, allowing their resolution through computational steps and algorithms. Evolving into a centering skill applicable across diverse professional domains, CT extends beyond computer science, reaching learners in various disciplines. Its impact on education is evident as teachers integrate it from preschool to higher education levels. The role and significance of CT in education can be outlined across four key dimensions, namely fostering problem-solving skills, preparing for the future workforce, fostering critical thinking skills, and exploring creativity. As the importance of CT continues to grow in the modern world, educators recognize the need to integrate CT into academic disciplines at all levels of education. However, the incorporation of CT into education presents both challenges and opportunities. One of the significant challenges educators face is how to design and deliver CT learning activities that align with their teaching goals and objectives. Teachers must have the knowledge and skills to design and implement practical CT learning activities and assess students' CT competencies. Addressing these challenges necessitates a robust foundation, where theoretical frameworks are crucial. Several theoretical frameworks have been proposed in the literature to address these challenges and guide educators in integrating CT into teaching. These frameworks offer guidelines and strategies for designing, delivering, and assessing CT learning activities. This chapter discusses three theoretical frameworks, namely problem-based instruction (PBI), Digital Storytelling (DST), and The CT Integration Elements. These frameworks were selected due to their alignment with the principles of design-based learning, which is the focus of this book. Each framework offers unique perspectives and approaches to integrating CT into teaching.

Keywords Computational thinking · Design-based learning · Integrating computational thinking · Theoretical background

---

T. K. Arslantaş (✉)
Department of Special Education, Aksaray University, Aksaray, Türkiye
e-mail: tugbakamaliarslantas@aksaray.edu.tr

M. Saritepeci and H. Yildiz Durak (eds.), *Integrating Computational Thinking Through Design-Based Learning*, https://doi.org/10.1007/978-981-96-0853-9_2

Introduction

Computational Thinking (CT), a cutting-edge term with roots dating back to the 1980s, has undergone significant evolution through the contributions of notable scholars (Bers, 2021; Denning, 2009; Papert, 1980; Wing, 2006, 2008). As argued by Papert (1980), the increasing use of computation in communication, research, culture, and business has led to the recognition of CT as essential for cognitive development. Since then, CT has evolved beyond its original description as a simple programming ability. Denning (2009) broadly characterized CT as the utilization of structured, algorithmic thinking to generate valuable outputs from inputs, while Wing (2006) emphasized that CT extends beyond technical nuances, positioning it as a life skill crucial for individuals. Wing (2008) further highlights CT's versatility, noting its potential in machine control while asserting that it is not exclusively dependent on machines. Building on Wing's (2006, 2008) studies, CT encompasses human behavior analysis, system design, and problem-solving techniques. Then, in simpler terms, CT is described as the thinking processes used to create problems in a way that their solutions can be shown through computational steps and algorithms (Aho, 2012). Despite the efforts of numerous scholars, achieving consensus on the concept of CT remains a challenge, reflecting the complex nature of this evolving intellectual domain. However, we can infer that, in essence, CT involves breaking down complex problems into smaller, more manageable ones that can be efficiently and automatedly solved.

In the forthcoming sections of this chapter, the critical function that CT plays in education, as well as the necessity, importance, and challenges that come with integrating CT into educational processes, are examined. The discussion will start by explaining the need for CT in education and highlighting why it has become a crucial component of learning. Following this, the role and importance of CT in education are explained, shedding light on the transformative impact it can have on students' cognitive development and problem-solving abilities. Subsequently, the chapter will focus on practical considerations, addressing the frameworks for effectively integrating CT into educational practices. Lastly, while acknowledging the nuances and potential obstacles that educators and institutions may meet, the difficulties that arise during this integration process will be covered. A comprehensive exploration of these aspects aims to provide insights and guidance for fostering a successful integration of CT into the educational landscape.

Computational Thinking in Education

CT is widely recognized as an essential skill, extending beyond the domain of computer science students to encompass learners across various professions. The cultivation of CT education necessitates the establishment of supportive learning environments, a notion endorsed by authoritative bodies (e.g., International Society for Technology in Education (ISTE) & Computer Science Teachers Association (CSTA), 2011) and underscored by researchers (e.g. Kordaki & Kakavas, 2017; Yildiz-Durak et al., 2023; Zhang & Nouri, 2019).

The ISTE and CSTA collaborative studies have significantly increased attention to CT research, particularly within K-12 educational contexts. Their joint project aims to enhance young learners' CT skills by illustrating how modern digital tools can address future challenges. The objective is to democratize CT education for K-12 educators, fostering an interest in thinking skills and integrating them into the curriculum, regardless of the presence of a computer. Emphasizing the development of these skills in diverse contexts, CT is defined by its problem-solving methods and interdisciplinary applications. The ultimate aspiration is for students to transition from mere tool users to proficient tool creators (Resnick et al., 2009), a competence beneficial both personally and academically. Notably, Grover and Pea (2013) emphasize the interdisciplinary nature of this approach, highlighting its positive impact on learning outcomes.

Just as definitions of CT vary, divergent perspectives exist regarding its integration into educational practices. Wing (2008) emphasizes how CT may be used to understand, analyze, and solve various issues using abstraction, algorithms, and systems thinking. Abstraction, a critical component, involves recognizing essential elements and their relationships while disregarding irrelevant distractions to effectively depict a problem. Algorithms, as abstract representations of processes, execute a series of operations using input from the problem representation to deliver a solution. Students can build larger systems by carefully utilizing abstraction and algorithms. This allows them to comprehend and deal with the complex relationships in real-world circumstances beyond the apparent stimulus.

Following research conducted by ISTE and CSTA, the integration of CT as a problem-solving methodology encompasses the following steps (2011):

- expressing problems in a way that makes it possible to use computers and other tools to solve them.
- employing algorithmic thinking, involving a systematic sequence of ordered steps, to automate solutions.
- arranging and evaluating data logically.
- using abstractions to represent data, such as simulations.
- finding, evaluating, and integrating potential solutions with the goal of reaching the most effective and efficient combination of actions and resources.
- applying this method of problem-solving in a general way to a range of issues.

Considering these perspectives, there is widespread agreement that individuals in all fields, regardless of their proficiency in programming, should have access to the science of computation (Yadav et al., 2016). Furthermore, it becomes evident that the central concept of CT in education lies in its application across diverse contexts and disciplines.

The Need for CT in Education

The emphasis on twenty-first-century skills has grown in the ever-changing educational landscape, acknowledging the dynamic demands of the modern world. Those skills are universal and comprise a large variety of skill sets that prepare learners for the future as expected future creators. The Partnership for twenty-first Century (P21) identifies three skill categories (Greenhill, 2010): (1) Literacy skills, such as media literacy, information, and communication technology literacy; (2) Learning skills, such as creativity, critical thinking, problem-solving, communication and collaboration; and (3) Life skills, such as adaptability and flexibility, initiative and self-direction, interpersonal and intercultural skills, productivity and accountability, and leadership and responsibility. The changing environment requires a comprehensive skill set beyond traditional academic abilities. This has prompted a reevaluation of educational priorities and methods.

Within this educational paradigm, CT is a keystone that effortlessly integrates and enhances critical twenty-first-century skills. The shift toward CT reflects a global acknowledgment of its significance as a fundamental skill for the next generation of students (Nouri et al., 2020; Weintrop et al., 2016). This acknowledgment underscores CT's essential role in equipping individuals to navigate the complexities of the modern world. Scholars widely advocate for including CT abilities in all curricula, emphasizing their role in preparing students with the necessary skills for the digital age (Gretter & Yadav, 2016).

The Role and Importance of CT in Education

This transformative shift positions CT at the forefront of educational innovation, aligning with the imperative development of twenty-first-century skills. The impact of CT is evident in its potential to enhance the development of cross-domain skills, ranging from literacy to problem-solving and self-regulation (Wright et al., 2012), as well as its application in addressing challenging tasks in science and math (Hambrusch et al., 2009). The role and significance of CT in education can be outlined across four key dimensions. The initial dimension underscores its pivotal role in fostering problem-solving skills (Shute et al., 2017). Beyond being a beneficial cognitive tool, CT is now considered indispensable, a robust method for effective problem-solving. ISTE (2011) defines CT as a problem-solving approach that enhances the combination of technology and thinking, aligning with Wing's (2008) perspective that it is a form of analytical thinking. According to Wing, CT is closely related to problem-solving and includes a variety of thinking techniques that should be smoothly incorporated into each child's educational path. CT's influence extends to nurturing competencies like problem-solving and creative thinking, which is crucial in shaping students' skills (Lye & Koh, 2014). Recent literature posits CT as an indispensable skill for individuals in the twenty-first century, designed to enhance problem-solving capabilities by leveraging principles from computer science (Selby, 2015). Functioning as a holistic problem-solving approach, CT engages cognitive functions linked to executive functions and, consequently, self-regulation (Myers, 2021). Furthermore, scholars have proposed that CT includes decomposing complex problems into simpler ones, creating algorithmic solutions, and simplifying difficulties to quickly develop heuristics (Barr & Stephenson, 2011). This perspective underscores the comprehensive nature of CT, positioning it as a cornerstone for fostering problem-solving skills in educational contexts.

The second dimension underscores the preparation for the future workforce. To meet the complex needs of the modern world, individuals must practice critical thinking, skillful communication, teamwork, and creative problem-solving (Chumark & Puncreobutr, 2016). This necessitates preparing individuals for competitive work environments in the future. All the previously discussed aspects highlight the significant influence of CT in shaping learners' futures, encompassing both academic readiness and workforce competence. To clarify, CT spans beyond computer science, extending to domain-specific problem-solving that aligns with twenty-first century skills such as digital literacy (Montuori et al., 2024). Early exposure to this powerful approach is imperative, significantly enhancing students' preparedness and capability to address future challenges effectively (Mihm, 2021). Even, Bers (2020) asserted that starting CT since from kindergarten is necessary for building future citizenry. Incorporating CT skills into a cohesive curriculum, as opposed to one that separates coding from social, emotional, and cognitive processing skills, can develop individuals who are more balanced and prepared for the challenges of the twenty-first century (Myers, 2021). Thus, introducing CT early builds a strong foundation, giving learners the skills and adaptability needed for the changing demands of future school and work.

The third dimension emphasizes the role of CT in fostering critical thinking skills. Notably, there are significant connections between CT, higher-order thinking, and executive functioning skills (Arfé et al., 2019), highlighting their focus on supporting reasoning and critical thinking (Yıldız-Durak & Saritepeci, 2018). ISTE (2015) expands the definition of CT to encompass creativity, algorithmic thinking, critical thinking, problem-solving, communication, and cooperation. In navigating the complex landscape of computational challenges, these comprehensive elements of CT necessitate the application of critical thinking skills, such as logical thinking, systems thinking, and assessing the evidence (Binkley et al., 2012), which are crucial for addressing complex computational problems (Li et al., 2023). As evident, CT education not only enhances children's CT skills but also improves non-cognitive skills like critical thinking (Su & Yang, 2023; Sung et al., 2017).

The fourth dimension emphasizes the exploration of creativity, a dynamic area of research that has attracted considerable attention, especially in the past decade. Early investigations primarily demonstrated an association between creativity and problem-solving on computerized programming platforms (Liu & Lu, 2002). Recent years have witnessed a shift in focus towards digital learning platforms, emphasizing CT, offering unique opportunities to include creative expression, and fostering innovative thinking. Yadav and Cooper (2017) emphasize the significance of these

CT fundamentals, such as problem-solving, abstraction, algorithmic thinking, and pattern recognition, in fostering creativity. Various studies conducted in diverse settings have revealed a reciprocal relationship between CT and creativity (Israel-Fishelson & Hershkovitz, 2022). As individuals engage with CT principles, they develop technical proficiency and cultivate a mindset that is inherently advantageous to generating novel ideas, exploring innovative solutions, and approaching challenges with a creative perspective.

Challenges of Integrating CT into Education

While acknowledging the significant role of CT as a valuable cognitive tool crucial for empowering young learners in navigating the complexities of the twenty-first century, the literature emphasizes various challenges related to its integration. These challenges include different dimensions, and the critical issues discussed here depend on the context and target groups. A primary challenge is the lack of a universally accepted definition for CT, with many countries still incorporating it into educational curricula at all levels (Lockwood & Mooney, 2018). This integration poses numerous challenges that extend beyond theoretical considerations.

From a broader perspective, two key areas demand further attention in CT integration. The first area concerns practical complexities, identifying effective activities and methodologies for integrating CT across diverse educational contexts (Tang et al., 2020). Incorporating CT into the curriculum proves challenging for educators and students, requiring the acquisition of new literacy skills alongside disciplinary knowledge (Lee et al., 2020). A significant challenge in integrating CT into regular school education is that many schools have yet to adopt it, often offering it as an optional course or an after-school activity, limiting its accessibility (Buffum et al., 2016; Vakil, 2018). Literature underlines teachers' challenges, including insufficient proficiency in CT concepts, knowledge of effective pedagogies, and strategies for incorporating CT into their primary academic subjects (Israel et al., 2022). This results in difficulties in selecting suitable learning tools and creating age-appropriate activities for young learners (Su & Yang, 2023), with challenges varying depending on the developmental level of countries, including issues like a lack of infrastructure and teacher qualifications.

The second aspect involves the development of suitable assessment frameworks for CT in novel settings (Tang et al., 2020). Existing systematic reviews highlight the immediate need to address uncertainties in optimal methods for assessing CT skills (Hsu et al., 2018; Su & Yang, 2023). Challenges include the necessity for valid and reliable CT assessments for children across a broader age range. The literature points out a lack of methods to assess students' application of learned skills to new problems and a general understanding of assessments in the context of CT integration (Israel et al., 2022).

These challenges are interconnected with the need to enhance the skills of existing teachers, provide training for new teachers, and address curriculum and resource issues (Sentance et al., 2013). Successful implementation of any instructional innovation, including CT, relies on teachers' preparation for integration. Despite these challenges, it is crucial to note that CT has made several contributions, and resolving these issues is imperative for educators to deliver integrated CT instruction aligned with contemporary education demands.

Theoretical Frameworks for Integrating CT into Education

While various studies highlight the effectiveness of CT in various aspects of student learning, the challenge lies in practically integrating CT into educational processes. Despite the recognized benefits, especially for students, the methods for integrating CT, particularly in K-12 education, remain vague to educators and researchers (Hsu et al., 2018). There is a crucial need for educators, practitioners, and teachers to grasp the complexity of the integration process. However, the primary focus should be on finding effective strategies for promoting successful CT learning activities (Denning, 2017).

This chapter aims to fill this knowledge gap by shedding light on the most employed strategies for integrating CT into educational processes. In doing so, it aims to provide valuable insights that can benefit a broad audience, offering practical guidance on effectively incorporating CT into diverse educational settings. The goal is to equip educators with the necessary tools and knowledge to navigate the challenges associated with CT integration, ultimately enhancing the overall educational experience for students. This section addresses three frameworks to integrating CT into educational processes, although it is acknowledged that various methods may exist.

These frameworks were selected due to their alignment with the principles of design-based learning (DBL), which is the focus of this book. DBL emphasizes hands-on, iterative processes that are integral to fostering CT skills. Each framework offers unique perspectives and approaches to integrating CT into teaching, thereby providing educators with comprehensive guidelines and strategies for designing, delivering, and assessing CT learning activities within a DBL context..

Problem-Based Instruction

Problem-based instruction (PBI) is the most frequently used strategy to integrate CT into education (e.g. Brush et al., 2020; Hsu et al., 2018; Kalelioglu et al., 2016; Kwon et al., 2021, 2023; Ma et al., 2021) possibly rooted in the definition of CT as a problem-solving approach by ISTE and CSTA in 2011. This method revolves around defining problems, utilizing computers or alternative tools, systematically organizing, and assessing data, using algorithmic thinking to automate solutions, achieving efficient resolutions, and applying generalizations to address other problems (Hsu et al., 2018).

The core concept in PBI for incorporating CT is to focus on specific problems formed by researchers or teachers. This approach allows students to engage in hands-on learning, enhancing their practice of CT, and developing content knowledge and other several skills while actively addressing these targeted problems. In that manner, PBI aligns well with the principles of DBL which emphasize iterative, hands-on processes that actively engage learners in solving real-world problems. Students are immersed in real-world contexts, making learning relevant and meaningful. PBI is also fed by DBL, as it focuses on solving problems, with DBL used in the problem-solving phase. PBI fosters critical thinking and problem-solving skills by requiring students to analyze and address complex issues and tackle design challenges that demand innovative solutions.

In essence, PBI is an experiential learning approach that allows students to acquire knowledge by actively engaging with real-world problems. It differs from traditional learning methods by employing diverse activities that enable students to qualify themselves while exploring and addressing identified problems. The primary objective is to allow students to acquire fresh insights and competencies within a supportive learning environment directly applicable to their future professional practice. Additionally, PBI yields superior outcomes in fostering students' CT compared to traditional lecturing methods (Tiwari et al., 2006). According to Wood (2003), students who engage in PBI identify their learning objectives in relation to real-world issues, investigate solutions on their own, then present their findings and suggestions to the group. In their study, Kwon et al. (2021) integrated PBI into CT processes for elementary students and investigated the effects of PBI on students' knowledge and attitudes towards computer science. Their findings suggested that PBI is an effective pedagogical strategy for implementing CT, as the participants significantly improved their CT knowledge. Similarly, Kwon et al. (2023) investigated the impact of a problem-based CT curriculum on elementary students who developed Scratch projects and indicated that students improved practicality in CT.

Essential components of CT are smoothly included in this method. Students use decomposition to break down complex problems into manageable parts, and as they create and improve step-by-step strategies, their algorithmic thinking skills are strengthened. As students go through various issues, their ability to recognize patterns is developed, and as they take critical information from real-world situations, abstraction naturally occurs. Furthermore, since PBI is iterative, students are encouraged to test and improve their solutions in response to feedback, parallel to the continuous iteration and development that characterizes CT. As a result, PBI enhances students' comprehension of the subject matter and the more extensive problem-solving process by fostering a solid foundation in CT abilities.

In their study, Kalelioglu et al. (2016) introduced a framework to serve as a guide for integrating CT into education through PBI. The framework comprises five key steps:

1. **Identify the Problem**: The initial step involves clearly defining the problem that students will be addressing within the context of the instructional material.

2. **Gathering, Representing, and Analyzing Data**: Once the problem is identified, students engage in activities related to collecting, representing, and analyzing relevant data.
3. **Generate, Select, and Plan Solutions**: In this phase, students are encouraged to creatively generate potential solutions, select the most viable options, and develop a comprehensive plan for implementing these solutions.
4. **Implement Solutions**: The focus shifts to putting the chosen solutions into action. Students actively implement their plans and apply CT strategies to address the identified problem.
5. **Assessing Solutions and Continuing for Improvement**: The final step involves evaluating the effectiveness of the implemented solutions. Students assess the outcomes, identify areas for improvement, and iterate on their solutions, fostering a continuous improvement cycle.

While the framework proposed by Kalelioglu et al. (2016) provides a structured guide for integrating CT through PBI, it is crucial to highlight the need for empirical evidence to support its effectiveness. Although the framework can potentially benefit those aiming to incorporate CT through PBI, empirical studies and practical implementations are necessary to validate its impact on educational outcomes. Similar to the framework proposed by Kalelioglu et al. (2016), another framework implemented in the studies is the IGGIA framework, which combines problem-solving and CT. This framework shares similar steps with Kalelioglu's model, with IGGIA representing Identifying the problem (I), gathering, representing, and analyzing data (G), generating, selecting, and planning solutions (G), implementing solutions (I), and assessing solutions and continuous improvement (A). In a study by Ma et al. (2021), the IGGIA framework was applied to primary school students working on Scratch projects. The researchers indicated that the IGGIA framework is efficacious in improving CT skills among primary school students and enhancing their self-efficacy regarding critical thinking and problem-solving.

Furthermore, it's essential to consider the limitations of PBI, mainly when applied to young children without prior experience in CT. PBI may challenge young learners to grasp abstract concepts and apply them effectively (Statter & Armoni, 2020). Therefore, special attention is recommended when implementing PBI with this demographic, and alternative strategies may need to be explored to cater to their developmental needs and capabilities.

Digital Storytelling

Digital storytelling (DST), an educational technique, blends the timeless art of storytelling with digital technologies to produce narratives based on a combination of various artifacts. In DST, the pedagogical value lies more in the process of creation rather than the final product, despite the narratives being the ultimate result of this attempt. It is important to note that DST has become an effective teaching and learning tool that involves students and teachers in real-world projects. While there are numerous studies in the literature investigating the PBI for integrating CT, relatively few studies emphasized the incorporation of CT within DST (Kordaki & Kakavas, 2017; Parsazadeh et al., 2021; Saritepeci, 2020; Yeni et al., 2022). The consistent theme across these studies is that combining DST with CT activities establishes a connection between the classroom and the real world, making the learning experience more enjoyable and significantly boosting student motivation (Parsazadeh et al., 2021).

DST is a method of integrating CT through a DBL perspective, significantly impacting the teaching of CT and demonstrating its application (Saritepeci, 2020). In the realm of computational literacy, DBL fosters creative CT, enabling students to develop and express their ideas through meaningful projects (Kim & Han, 2012). From this perspective, DST encourages students to think critically and solve problems creatively by designing and refining their narratives or solutions iteratively.

Kordaki and Kakavas (2017) conducted an initial study, establishing DST as a framework for integrating CT in K-12 education. Their primary aim was to endorse DST as an educational framework to enhance CT skills among K-12 level students. They outlined four stages in DST, each linked with associated CT skills.

1. **Setting the Stage**: Students analyze a given DST topic, choose a sub-topic, and consider the target audience and contextual factors, fostering critical and logical thinking.
2. **Design of the Story**: Research skills are employed to deepen knowledge on the chosen sub-topic. Scriptwriting and storyboarding involve logical thinking, problem decomposition, and aesthetic skills.
3. **Digital Story Development**: In the production stage, students organize and combine files, adding titles, narratives, and multimedia elements. This process demonstrates aesthetic skills, critical thinking, and data representation.
4. **Assessment of the Digital Story**: Evaluation criteria are provided during story production, utilizing a rubric for self-assessment, peer evaluation, and assessing other teams' stories. Evaluation enhances critical thinking and debugging skills.

Yeni et al. (2022) conducted a study that implemented DST to integrate CT utilizing the framework proposed by Kordaki and Kakavas (2017). Their findings suggested that integrating CT through DST in language learning environments holds promising potential as a motivating learning mechanism. Parsazadeh et al. (2021) also investigated DST with CT among elementary school students and suggested that integrating DST with CT as a strategy increased the motivation and performance of students in language learning. Saritepeci (2020) conducted an experimental study investigating the effect of design-based and programming activities on CT. At the core of these design-based activities, he used DST and found that design-based learning activities contributed to participants' creativity and problem-solving skills. While the conclusions drawn are promising, it is essential to recognize the necessity for additional studies to thoroughly elucidate the role of DST as a practical framework. Notably, the framework's development and testing primarily occurred in K-12 settings, underscoring the need for further research.

The CT Integration Elements

PBI and DST, as mentioned earlier, are general educational frameworks adapted for the integration of CT. In contrast to these frameworks, another study in the literature has developed initially a disciplinary-based framework specifically for CT. This particular study, initiated by Malyn-Smith et al. (2018), aims to provide a comprehensive framework for CT within the context of a specific discipline. Specifically, they aimed to explore methods for promoting, instructing, and applying CT within various academic subjects in primary and secondary education. It establishes an initial collection of "elements" delineating CT practices that connect learning with the demands of advanced STEM settings. These CT integration elements extend beyond traditional programming activities, serving as a bridge between CT in K-12 classrooms and professional practices.

Similar to PBI and DST, this framework inherently reflects a DBL perspective. As outlined in the following guide, the elements of CT Integration encompass design thinking and processes, which align with DBL principles. This approach supports a deeper understanding of complex problems and fosters skills such as creativity, critical thinking, and collaboration, essential for effective educational practices.

The study proposes a comprehensive guide for integrating CT with the following key elements (Lee & Malyn-Smith, 2020).

1. **Understand (Complex) Systems**: Using CT modeling interactions in complex systems helps reveal emergent patterns. Computer modeling and simulation enable students to observe how complex systems behave under different conditions. Students can raise inquiries beginning with "what if?" and carry out scientific experiments to find solutions.
2. **Innovate with Computational Representations**: CT facilitates the design and development of innovations through conceptualizing and developing ideas as computations. Students can be introduced to computational representations, such as understanding colors as RGB values.
3. **Design Solutions that Leverage Computational Power/Resources**: CT aids scientists in designing solutions that optimize computational resources and power for efficient problem-solving. Within K-12 environments, teachers can inspire students to consider alternative problem-solving approaches when faced with a larger input set. For instance, instead of arranging 10 items, how might students approach the task of sorting 10,000 items?
4. **Engage in Collective Sense-Making Around Data**: CT involves analyzing data collectively, uncovering patterns, and visualizing multidimensional data to enhance understanding. In K-12 environments, educators can request small student groups to conduct simulations using a portion of the inputs, followed by sharing their output data and analyses. The compilation and analysis of the collective data demonstrate how each segment contributes to comprehending the entire dataset.
5. **Understand Potential Consequences of Actions**: Scientists use CT to simulate and predict outcomes and explore the space of possible inputs. Within

K-12 educational settings, students can grasp how cause-and-effect connections are employed to anticipate results. Additionally, students can initiate an understanding of the range of inputs by parameterizing models.

Integrating Computational Thinking in the Age of AI

AI technologies offer robust support for students in applying their knowledge, fostering creativity, and enhancing critical thinking skills. By incorporating AI applications into educational settings, students can better develop the competencies needed to tackle design-based challenges. AI facilitates this by providing tools and resources that help students engage with algorithmic problem-solving, pattern recognition, and logical reasoning. Through interactive and adaptive AI-driven exercises, students are guided in solving complex problems more effectively, which in turn improves their CT skills. This integration allows for a more nuanced approach to learning, where AI not only aids in understanding and application but also actively enhances students' abilities to think critically and logically, preparing them for future problem-solving tasks.

Conclusion

This chapter focused on theoretical foundations, practical considerations, and challenges associated with integrating CT into education. As a term that has evolved over decades, CT is recognized not just as a programming skill but as a fundamental cognitive tool crucial for navigating the complexities of the modern world. The chapter emphasizes the need for a comprehensive understanding of CT and its applications across diverse educational contexts.

The role and importance of CT in education are explored across various dimensions, highlighting its transformative impact on problem-solving, future workforce preparation, critical thinking, and creativity. The literature review underscores that CT is not confined to computer science but spans across disciplines, making it an essential skill for all learners.

The subsequent section focuses on practical strategies for integrating CT into educational processes. Three frameworks are discussed, including PBI, DST, and The CT Integration Elements, a disciplinary-based framework proposed by Malyn-Smith et al. (2018). Each framework offers unique perspectives and approaches to fostering CT skills, catering to the diverse needs of learners. The chapter recognizes the significance of empirical evidence to support the effectiveness of these frameworks and emphasizes the importance of addressing the challenges associated with CT integration.

However, challenges persist, including the lack of a universally accepted definition for CT, practical complexities in integrating CT across diverse contexts, and the need for suitable assessment frameworks. The chapter highlights the interconnected nature of these challenges and underscores the importance of addressing issues related to teacher proficiency, curriculum development, and resource availability.

In essence, while acknowledging the problems, the chapter advocates for incorporating CT in education, presenting it as an indispensable tool for empowering learners in the twenty-first century. It calls for collaborative efforts from educators, researchers, and policymakers to overcome challenges and create a conducive environment for the seamless integration of CT into educational practices. Ultimately, the goal is to equip learners with the skills and mindset necessary to succeed in a growing digital landscape.

Summary of Key Points

This section is dedicated to the theoretical framework for integrating CT in education.

Key points:

- The shift toward CT reflects a global acknowledgment of its significance as a fundamental skill for the next generation of students. CT is a crucial skill that involves problem-solving, algorithmic thinking, and computational processes. It is increasingly recognized as essential across various disciplines and educational levels.
- Integrating CT into education offers several benefits, including fostering problem-solving skills, preparing students for the future workforce, developing critical thinking abilities, and encouraging creativity.
- Integrating CT into teaching practices presents challenges as well. Educators need effective frameworks to guide CT integration, design learning activities, and assess student outcomes.
- The chapter focuses on three theoretical frameworks that align with DBL principles, which emphasize student engagement and problem-solving:
  - **Problem-Based Instruction (PBI)**: Emphasizes real-world problem-solving and student-centered learning.
  - **Digital Storytelling (DST)**: Combines storytelling with technology to promote creativity and critical thinking.
  - **CT Integration Elements**: Provides a structured approach to integrating CT into different subject areas.

    By understanding and applying these frameworks, educators can effectively integrate CT into their teaching practices and equip students with the necessary skills for the twenty-first century.

Implications for Future Practice and Research

All the issues discussed in this chapter provide insights into the benefits, challenges, and theoretical frameworks for effective CT integration. Given the significance of CT in the education system and considering ongoing technological advancements, it is clear that CT will continue to gain importance. DBL shares several common points with CT integration, impacting the integration processes. Thus, the integration of CT into education requires continued efforts in curriculum development, teacher training, and resource allocation to ensure comprehensive and equitable implementation across diverse educational contexts.

Future research should focus on validating the effectiveness of CT integration frameworks, refining its definition, and exploring innovative pedagogical approaches to enhance CT education. While the chapter highlights various frameworks for CT integration, further empirical research is needed to validate their effectiveness across different educational contexts. Longitudinal studies should be conducted to assess the long-term impact of CT education on students' problem-solving abilities, critical thinking, and creativity.

References

Aho, A. V. (2012). Computation and computational thinking. *The Computer Journal, 55*(7), 832–835. https://doi.org/10.1093/comjnl/bxs074

Arfé, B., Vardanega, T., Montuori, C., & Lavanga, M. (2019). Coding in primary grades boosts children's executive functions. *Frontiers in Psychology, 10*, 2713. https://doi.org/10.3389/fpsyg.2019.02713

Barr, V., & Stephenson, C. (2011). Bringing computational thinking to K-12: What is involved and what is the role of the computer science education community? *ACM Inroads, 2*(1), 48–54. https://doi.org/10.1145/1929887.1929905

Bers, M. (2020). Playgrounds and microworlds: Learning to code in early childhood. In N. Holbert, M. Berland, & Y. B. Kafai (Eds.), *Designing constructionist futures: The art, theory and practice of learning designs* (pp. 31–38). The MIT Press. https://doi.org/10.7551/mitpress/12091.003.0006

Bers, M. (Ed.). (2021). *Teaching computational thinking and coding to young children.* IGI Global. https://doi.org/10.4018/978-1-7998-7308-2

Binkley, M., Erstad, O., Herman, J., Raizen, S., Ripley, M., Miller-Ricci, M., & Rumble, M. (2012). Defining twenty-first century skills. In P. Griffin, B. McGaw, & E. Care (Eds.), Assessment and teaching of 21st century skills (pp. 17–66). Springer https://doi.org/10.1007/978-94-007-2324-5_2

Brush, T., Ottenbreit-Leftwich, A., Kwon, K., & Karlin, M. (2020). Implementing socially relevant problem-based computer science curriculum at the elementary level: Students' computer science knowledge and teachers' implementation needs. *Journal of Computers in Mathematics and Science Teaching, 39*(2), 109–123.

Buffum, P. S., Frankosky, M. H., Boyer, K. E., Wiebe, E. N., Mott, B. W., & Lester, J. C. (2016, February). Empowering all students: Closing the computer science confidence gap with an in-school initiative for middle school students. In *Proceedings of the 47th ACM technical symposium on computing science education* (pp. 382–387).

Chumark, C., & Puncreobutr, V. (2016). Developing basic mathematical skills of pre-school children by using plasticized clay. *Journal of Education and Practice, 7*(12), 180–183.
Denning, P. J. (2009). The profession of IT beyond computational thinking. *Communications of the ACM, 52*(6), 28–30. https://doi.org/10.1145/1516046.1516054
Denning, P. J. (2017). Remaining trouble spots with computational thinking. *Communications of the ACM, 60*(6), 33–39. https://doi.org/10.1145/2998438
Greenhill, V. (2010). 21st century knowledge and skills in educator preparation. *Partnership for 21st century skills.* https://eric.ed.gov/?id=ED519336
Gretter, S., & Yadav, A. (2016). Computational thinking and media & information literacy: An integrated approach to teaching twenty-first century skills. *TechTrends, 60*, 510–516. https://doi.org/10.1007/s11528-016-0098-4
Grover, S., & Pea, R. (2013). Computational thinking in K–12: A review of the state of the field. *Educational Researcher, 42*(1), 38–43. https://doi.org/10.3102/0013189X12463051
Hambrusch, S., Hoffmann, C., Korb, J. T., Haugan, M., & Hosking, A. L. (2009). A multidisciplinary approach towards computational thinking for science majors. *ACM Sigcse Bulletin, 41*(1), 183–187. https://doi.org/10.1145/1539024.1508931
Hsu, T. C., Chang, S. C., & Hung, Y. T. (2018). How to learn and how to teach computational thinking: Suggestions based on a review of the literature. *Computers & Education, 126*, 296–310. https://doi.org/10.1016/j.compedu.2018.07.004
Israel-Fishelson, R., & Hershkovitz, A. (2022). Studying interrelations of computational thinking and creativity: A scoping review (2011–2020). *Computers & Education, 176*, 104353. https://doi.org/10.1016/j.compedu.2021.104353
Israel, M., Liu, R., Yan, W., Sherwood, H., Martin, W., Fancsali, C., Rivera-Cash, E., & Adair, A. (2022). Understanding barriers to school-wide computational thinking integration at the elementary grades: Lessons from three schools. In *Computational thinking in PreK-5: Empirical evidence for integration and future directions* (pp. 64–71).
ISTE. (2011). *Computational thinking in K-12 education leadership toolkits.* https://www.iste.org/resources/attachmentdownload?ID3413
ISTE. (2015). *ISTE standards for students.* https://www.iste.org/docs/pdfs/20-14_ISTE_Standards-S_PDF.pdf
ISTE & CSTA. (2011). *Operational definition of computational thinking for K-12 education.* http://www.iste.org/docs/ct-documents/computational-thinking-operational-definition-flyer.pdf
Kalelioglu, F., Gülbahar, Y., & Kukul, V. (2016). A framework for computational thinking based on a systematic research review. *Baltic Journal of Modern Computing, 4*(3), 583–596.
Kordaki, M., & Kakavas, P. (2017). Digital storytelling as an effective framework for the development of computational thinking skills. In *Edulearn17 proceedings* (pp. 6325–6335). IATED.
Kwon, K., Jeon, M., Guo, M., Yan, G., Kim, J., Ottenbreit-Leftwich, A. T., & Brush, T. A. (2023). Computational thinking practices: Lessons learned from a problem-based curriculum in primary education. *Journal of Research on Technology in Education, 55*(4), 590–607. https://doi.org/10.1080/15391523.2021.2014372
Kwon, K., Ottenbreit-Leftwich, A. T., Brush, T. A., Jeon, M., & Yan, G. (2021). Integration of problem-based learning in elementary computer science education: Effects on computational thinking and attitudes. *Educational Technology Research and Development, 69*, 2761–2787. https://doi.org/10.1007/s11423-021-10034-3
Lee, I., Grover, S., Martin, F., Pillai, S., & Malyn-Smith, J. (2020). Computational thinking from a disciplinary perspective: Integrating computational thinking in K-12 science, technology, engineering, and mathematics education. *Journal of Science Education and Technology, 29*, 1–8.
Lee, I., & Malyn-Smith, J. (2020). Computational thinking integration patterns along the framework defining computational thinking from a disciplinary perspective. *Journal of Science Education and Technology, 29*, 9–18. https://doi.org/10.1007/s10956-019-09802-x

Li, W., Liu, C. Y., & Tseng, J. C. (2023). Effects of the interaction between metacognition teaching and students' learning achievement on students' computational thinking, critical thinking, and metacognition in collaborative programming learning. *Education and Information Technologies, 28*, 12919–12943. https://doi.org/10.1007/s10639-023-11671-2

Liu M. C., & Lu H. F. (2002, August 18–21). A study on the creative problem-solving process in computer programming. In *Proceedings of the ınternational conference on engineering education, Manchester, UK*. ICEE.

Lockwood, J., & Mooney, A. (2018). developing a computational thinking test using Bebras problems. In *Joint proceedings of the CC-TEL 2018 and TACKLE 2018 workshops, co-located with 13th European conference on technology enhanced learning (EC-TEL 2018).*

Lye, S. Y., & Koh, J. H. L. (2014). Review on teaching and learning of computational thinking through programming: What is next for K-12? *Computers in Human Behavior, 41*, 51–61. https://doi.org/10.1016/j.chb.2014.09.012

Ma, H., Zhao, M., Wang, H., Wan, X., Cavanaugh, T. W., & Liu, J. (2021). Promoting pupils' computational thinking skills and self-efficacy: A problem-solving instructional approach. *Educational Technology Research and Development, 69*(3), 1599–1616. https://doi.org/10.1007/s11423-021-10016-5

Malyn-Smith, J., Lee, I., Martin, F. G., Grover, S., Evans, M. A., & Pillai, S. (2018). Developing a framework for computational thinking from a disciplinary perspective. In S. C. Kong, D. Andone, G. Biswas, T. Crick, H. U. Hoppe, T. C. Hsu, & J. Vahrenhold (Eds.), *Conference proceedings of the ınternational conference on computational thinking education 2018* (pp. 182–186). The Education University of Hong Kong.

Mihm, C. M. (2021). Why teach coding to early elementary learners. In *Teaching computational thinking and coding to young children* (pp. 21–40). IGI Global.

Montuori, C., Gambarota, F., Altoé, G., & Arfé, B. (2024). The cognitive effects of computational thinking: A systematic review and meta-analytic study. *Computers & Education, 210*, 104961. https://doi.org/10.1016/j.compedu.2023.104961

Myers, E. K. (2021). The role of executive function and self-regulation in the development of computational thinking. In M. Bers (Ed.), *Teaching computational thinking and coding to young children* (pp. 64–83). IGI Global.

Nouri, J., Zhang, L., Mannila, L., & Norén, E. (2020). Development of computational thinking, digital competence and 21st century skills when learning programming in K-9. *Education Inquiry, 11*(1), 1–17. https://doi.org/10.1080/20004508.2019.1627844

Papert, S. (1980). *Mindstorms: Children, computers, and powerful ideas.* Basic Books.

Parsazadeh, N., Cheng, P. Y., Wu, T. T., & Huang, Y. M. (2021). Integrating computational thinking concept into digital storytelling to improve learners' motivation and performance. *Journal of Educational Computing Research, 59*(3), 470–495. https://doi.org/10.1177/0735633120967

Resnick, M., Maloney, J., Monroy-Hernández, A., Rusk, N., Eastmond, E., Brennan, K., Millner, A., Rosenbaum, E., Silver, J., Silverman, B., & Kafai, Y. (2009). Scratch: Programming for all. *Communications of the ACM, 52*(11), 60–67.

Saritepeci, M. (2020). Developing computational thinking skills of high school students: Design-based learning activities and programming tasks. *The Asia-Pacific Education Researcher, 29*(1), 35–54. https://doi.org/10.1007/s40299-019-00480-2

Selby, C. C. (2015). Relationships: Computational thinking, pedagogy of programming, and Bloom's Taxonomy. In *Proceedings of the workshop in primary and secondary computing education* (pp. 80–87). ACM. https://doi.org/10.1145/2818314.2818315

Sentance, S., Dorling, M., & McNicol, A. (2013). Computer science in secondary schools in the UK: Ways to empower teachers. In *International conference on ınformatics in schools: situation, evolution, and perspectives* (pp. 15–30). Springer Berlin Heidelberg.

Shute, V. J., Sun, C., & Asbell- Clarke, J. (2017). Demystifying computational thinking. *Educational Research Review, 22*, 142–158. https://doi.org/10.1016/j.edurev.2017.09.003

Statter, D., & Armoni, M. (2020). Teaching abstraction in computer science to 7th grade students. *ACM Transction on Computing Education, 20*(1), 1–37. https://doi.org/10.1145/3372143

Su, J. & Yang, W. (2023). A systematic review of integrating computational thinking in early childhood education. *Computers and Education Open*, 100122. https://doi.org/10.1016/j.caeo.2023.100122

Sung, W., Ahn, J., & Black, J. B. (2017). Introducing computational thinking to young learners: Practicing computational perspectives through embodiment in mathematics education. *Technology, Knowledge and Learning, 22*, 443–463. https://doi.org/10.1007/s10758-017-9328-x

Tang, X., Yin, Y., Lin, Q., Hadad, R., & Zhai, X. (2020). Assessing computational thinking: A systematic review of empirical studies. *Computers & Education, 148*, 103798. https://doi.org/10.1016/j.compedu.2019.103798

Tiwari, A., Chan, S., Wong, E., Wong, D., Chui, C., Wong, A., & Patil, N. (2006). The effect of problem-based learning on students' approaches to learning in the context of clinical nursing education. *Nurse Education Today, 26*(5), 430–438.

Vakil, S. (2018). Ethics, identity, and political vision: Toward a justice-centered approach to equity in computer science education. *Harvard Educational Review, 88*(1), 26–52. https://doi.org/10.17763/1943-5045-88.1.26

Weintrop, D., Beheshti, E., Horn, M., Orton, K., Jona, K., Trouille, L., & Wilensky, U. (2016). Defining computational thinking for mathematics and science classrooms. *Journal of Science Education and Technology, 25*, 127–147. https://doi.org/10.1007/s10956-015-9581-5

Wing, J. M. (2006). Computational thinking. *Communications of the ACM, 49*(3), 33–35. https://doi.org/10.1145/1118178.1118215

Wing, J. M. (2008). Computational thinking and thinking about computing. *Philosophical Transactions of the Royal Society A: Mathematical, Physical and Engineering Sciences, 366*(1881), 3717–3725.

Wood, D. F. (2003). Problem based learning. *BMJ, 326*(7384), 328–330.

Wright, G. A., Rich, P., & Leatham, K. R. (2012). How programming fits with technology education curriculum. *Technology and Engineering Teacher, 71*(7), 3.

Yadav, A., & Cooper, S. (2017). Fostering creativity through computing. *Communications of the ACM, 60*(2), 31–33. https://doi.org/10.1145/3029595

Yadav, A., Gretter, S., Hambrusch, S., & Sands, P. (2016). Expanding computer science education in schools: Understanding teacher experiences and challenges. *Computer Science Education*, 1–20. https://doi.org/10.1080/08993408.2016.1257418

Yeni, S., Nijenhuis-Voogt, J., Hermans, F., & Barendsen, E. (2022, October). An ıntegration of computational thinking and language arts: The contribution of digital storytelling to students' learning. In *Proceedings of the 17th workshop in primary and secondary computing education* (pp. 1–10).

Yıldız-Durak, H., & Saritepeci, M. (2018). Analysis of the relation between computational thinking skills and various variables with the structural equation model. *Computers & Education, 116*, 191–202. https://doi.org/10.1016/j.compedu.2017.09.004

Yildiz Durak, H., Saritepeci, M., & Durak, A. (2023). Modeling of relationship of personal and affective variables with computational thinking and programming. *Technology, Knowledge and Learning, 28*(1), 165–184. https://doi.org/10.1007/s10758-021-09565-8

Zhang, L., & Nouri, J. (2019). A systematic review of learning computational thinking through Scratch in K-9. *Computers & Education, 141*. https://doi.org/10.1016/j.compedu.2019.103607

Chapter 3
Design-Based Learning Approaches for Computational Thinking

Nilüfer Atman Uslu

Abstract Education systems intend to raise generations that produce digital solutions for our planet's problems by providing students with computational thinking (CT) skills through programming activities. To realize this vision, children need to understand the needs and requirements of the beneficiaries in the digital products they will develop and produce solutions to these problems. Accordingly, design-based learning (DBL) can provide a framework for teachers who want to lead students in acquiring CT skills. Accordingly, this chapter suggests educational implications by focusing on integrating CT and DBL. The chapter begins with an overview of design thinking (DT) and definitions of DBL. Also, the chapter highlights the five key ideas found in DBL's framework: (a) Authentic learning environment, (b) Learner-centered approach, (c) Inquiry and investigation, (d) Prototyping and iteration, (e) Fostering deep learning. Then, by touching on the disadvantages and advantages of DBL, the points that should be considered to expand its role in learning and teaching processes are emphasized. The chapter then summarizes the stages of design thinking (DT), providing an understanding of the methodological process that supports DBL. Following this introduction to DBL, examples of how it can help learners' CT development are included. Thus, examples of how CT and DT can intersect are highlighted, and the potential benefits of combining the two approaches are emphasized. In conclusion, this chapter highlights the transformative potential of combining DBL and CT in educational contexts.

Keywords Design thinking · Computational thinking · Design-based learning · Pedagogical implications

N. Atman Uslu (✉)
Department of Primary Education, Buca Faculty of Education, Dokuz Eylül University, İzmir, Turkey
e-mail: nilufer.atmanuslu@deu.edu.tr

M. Saritepeci and H. Yildiz Durak (eds.), *Integrating Computational Thinking Through Design-Based Learning*, https://doi.org/10.1007/978-981-96-0853-9_3

Design-Based Learning

DT is an iterative method that offers a framework for designing products that address the problems faced by living things. Through empathy with potential beneficiaries, members of the design team define the challenge and come to comprehend it. The group creates and learns about the challenge by developing empathy for possible beneficiaries. After this, concepts for the challenge are generated, tested, and revised, considering the feedback obtained. This process is iterative rather than linear. By applying this framework in educational settings, DBL effectively emphasizes design while fostering a problem-based learning approach (Gómez Puente et al., 2011). DBL can support children's creativity, and collaborative problem-solving. To realize this potential, students need to acquire skills such as dealing with complexity and approaching problems from different perspectives and points of view. Education systems intend to raise generations that produce digital solutions for our planet's problems by providing students with CT skills through programming activities. To realize this vision, children need to understand the needs and requirements of the beneficiaries in the digital products they will develop and produce solutions to these problems. CT includes problem decomposition, abstraction, algorithms, and testing. Accordingly, DT can provide a framework for teachers who want to lead students in acquiring CT skills. This chapter focuses on how DBL can promote CT skills in students.

Human-Centered Solutions: Bridging DT and DBL

Although DBL and DT differ in terms of focus, they are interrelated concepts. DT, also referred to as human-centered design, is an innovative approach to problem-solving that concentrates on creating user-centric, efficient solutions for clearly outlined issues (Grau & Rockett, 2022). The designer explores, reveals, and elucidates the uncertain phenomenon, proposes alternative options, and manipulates and reshapes the material (Ejsing-Duun & Skovbjerg, 2019). DT is a process in which multidisciplinary teams adopt a user-oriented approach to find solutions to complex problems (Carlgren et al., 2014). The practical applications of DT have been formalized as the innovation process begins with user engagement and a thorough exploration of human-related issues (Carella et al., 2023). DT differs from other innovation approaches because its human-centered aspect prioritizes people's satisfaction (Nawang, 2022). It involves a human-centered, action-oriented, and mindful process, focusing on hands-on projects that explore real problems with empathy and create feasible prototypes through creative ideation (Cha, 2019).

At this point, DBL is a self-directed approach where students learn by designing hands-on solutions that meet academic, societal, and industry standards, utilizing a problem-solving structure in project-based education (Chandrasekaran et al., 2013). DBL, rooted in the constructionist theory that emphasizes learners actively constructing knowledge, involves students applying theoretical knowledge to create artifacts or systems to address real-world challenges, providing a rich learning experience through the design process (Azizan & Abu Shamsi, 2022). Through the creation and execution of original, creative, and inventive projects, DBL enables students to generate cognitive concepts and gives them the freedom to begin the learning process according to their individual preferences, learning styles, and range of skill sets (Doppelt et al., 2008). Students can enhance their critical thinking abilities and develop creative solutions to problems in the real world by implementing DT concepts into DBL.

Human-centricity, inventive perspective, experiential learning, visual representation, holistic approach, and divergent and convergent thinking are the six essential DT concepts in creative problem-solving (Carella et al., 2023). Micheli et al. (2019) listed ten critical attributes of DT: creativity and innovation, user focus and engagement, problem-solving, iteration and experimentation, interdisciplinary cooperation, visualization ability, holistic approach, abductive reasoning, tolerance for failure and uncertainty, and combining analysis and intuition. This approach sets itself apart from problem-based learning by emphasizing empathy and iterative evaluation from a human-centered perspective through divergent and convergent thinking and encourages creative ideas (Cha, 2019).

Key Elements

This section includes the key elements that need to be considered to effectively implement DBL in learning and teaching processes.

Authentic Learning Environment

Authentic tasks based on real-life facts are one of the tenets of constructivist learning (Pande & Bharathi, 2020). Because authentic tasks involve real-world problems, they offer students a meaningful starting point for the steps they will follow in DBL. Students develop deep understanding through authentic tasks, increasing their engagement in learning and encouraging ownership and autonomy. As a result, an authentic learning environment is an essential component of DBL. Tasks that are authentic and incorporate DT encompass practical applications in the real world, such as chances for working together, merging various subject areas, evaluation that flows smoothly, refined end products that hold value independently, and permit different solutions and a variety of results (Inder & Withell, 2013). Having access to integrative and authentic learning environments allows students to creatively intervene to model, adapt, and develop ideas through interactive processes, thus becoming creative problem solvers and designers who can engage with emerging technologies in the future (Seitamaa-Hakkarainen, 2011).

Learner-Centered Approach

DT serves as a method for developing learner-centered initiatives and is beneficial in teaching students to enhance their problem-solving abilities and create an engaging student experience (Grau & Rockett, 2022). At the same time, DBL and student-centered learning focus on interaction and developing empathy with the user or student, encouraging participation in the social environment, and increasing awareness of understanding different people's experiences (Altay, 2014). Student-centered design places students at the core, considers learning outcomes, is flexible across curricular levels and contexts, and encourages collaboration and co-creation, fostering a bias towards action and embracing failure as a learning opportunity (Grau & Rockett, 2022). Student-centered learning is crucial in DBL because it prioritizes students' needs and experiences while addressing essential learning outcomes. Indeed, in DBL, students are at the center of the learning process and can gain problem-solving skills, creativity, and a deeper understanding of the real-world impacts of design decisions.

Inquiry and Investigation

Inquiry begins with doubt and questioning, creating a situation for testing hypotheses and understanding the problem (Ejsing-Duun & Skovbjerg, 2019). Inquiry supports students in scientifically exploring concepts and cause-effect relationships (Prayogi et al., 2023). The DT approach follows a different and non-linear path to problem-solving by creatively redefining and challenging existing problems, practices, and assumptions (Carella et al., 2023). Researchers have found that learning processes involving DBL positively impact students' creativity and problem-solving skills (Yalçın & Erden, 2021). In DBL, students ask questions, evaluate possibilities, formulate hypotheses, and seek solutions by testing them. Learning in this way naturally involves an iterative process of questioning and examining. DBL frames the necessary foundation as it is a systematic approach to creative problem-solving (Zhu et al., 2024).

Prototyping and Iteration

The prototype is a simple solution model designed to solve a problem so the design team can make the necessary corrections and refinements quickly and easily (Dam & Siang, 2020). The design team collects the prototype of a solution and the opinions of potential beneficiaries, gathering data on both successful and unsuccessful aspects. This way, the designer(s) can make necessary revisions before the solution goes live.

Of course, this involves an iterative development process, feedback, and reconsidering. Prototyping and iteration also help the design team understand how the target audience perceives the problem and solution they have in mind.

Upon receiving feedback, the design team may realize there was an error in formulating the problem. In this case, reframing the developed solution or model in question may be required. In this context, prototyping and iteration provide significant advantages in terms of time and economy by providing insight into the shortcomings and improvements of the solution or product before it is developed for real life.

Fostering Deep Learning

Deep learning involves learners critically reflecting on real-life scenarios, actively linking past and current knowledge, and tackling intricate problems through information processing. Deep learning aims to develop students' higher-order thinking and problem-solving capabilities, making it an effective way to enhance core qualities and likely to become a trend in curriculum reform (Zhang, 2020). Accordingly, DBL provides the context for students to gain profound learning experiences through fostering the development of higher-order thinking skills. Utilizing DBL engages students' advanced cognitive skills and abilities and embodies a crucial objective in the evolving learning process (Weng et al., 2023).

Benefits and Challenges

While DBL creates meaningful learning opportunities, it also requires challenging tasks. According to Wagner (2014), teachers think that DBL helps enable transdisciplinary practices and supports students' cooperation, motivation, and critical thinking skills. However, it needs help regarding assessment, time, and pedagogy. At the same time, students stated that DBL supported them in communication, collaboration, thinking from multiple perspectives, and gaining real-life experience and motivation while they needed help with critical thinking skills and time (Wagner, 2014). Pereira et al. (2021) examined engineers' benefit-challenge perceptions regarding DT in professional business with a mixed method. Research findings showed that the benefits of DT include increased collaboration, better understanding, identification of actual needs, increased creativity, empathy, and more efficient process management. The same study reported challenges such as lack of time, people engagement, lack of values, need for cultural change, limited knowledge, understanding of the problem, engagement in multidisciplinary teams, and low maturity to DT (Pereira et al., 2021). Hehn et al. (2018) summarize the main challenges related to DT in software development as follows:

- Strong focus on user requirements while neglecting software and system requirements.
- Weak links between needs, insights, learnings, and requirement
- Lack of formalizing context requirements.
- Lack of motivation to specify requirements systematically.
- Lack of time to specify requirements systematically.
- Knowledge is implicit.

Gonera and Pabst (2019) classified the possible enablers of DT as follows: (a) Experiential learning, (b) Change agent, (c) Advocate for DT, (d) Established set of DT tools and formats, (e) Team reflexivity.

In conclusion, DT is a powerful problem-solving approach with numerous benefits. It encourages creativity and innovation, allowing for out-of-the-box solutions to complex problems, and promotes empathy, as it requires designers to deeply understand the needs and perspectives of the end-users. Furthermore, DT encourages cooperation and teamwork and frequently entails interdisciplinary groups coming together to solve problems. However, there are drawbacks to DT as well. For people and organizations used to conventional problem-solving techniques, it can be challenging to adopt this new way of thinking.

Stages of DT

An iterative cycle of design and revision, reflective learning with feedback and modification, collaborative and integrative group work, and situational learning through inquiry activities are the hallmarks of the DBL approach (Weng et al., 2023). Within the framework of these elements, DBL has specific steps. There are several models for the steps in DT. Hasso-Plattner Institute of Design process (2016) guide recommends five steps for DT: (a) Empathy entails understanding people's needs, thoughts, and values within the context of a design issue. To create creative solutions based on insights derived from actually seeing and interacting with people, empathy is crucial to human-centered design. (b) The Define mode of the design process emphasizes that the goals are to make the design space more focused and precise, formulate a meaningful problem statement, and combine empathetic research with insightful analysis to address the design challenge. (c) Ideate stage is dedicated to producing ideas and concepts that provide the basis for creating prototypes and putting creative solutions into practice. (d) The Prototype mode involves creating iterative artifacts to address questions and move closer to the final solution, starting with low-resolution prototypes for early feedback and progressing to more refined prototypes as the project advances. (e) The test mode involves gathering user feedback on prototypes, gaining empathy, understanding the user and the problem, and focusing on learning and refinement rather than simply asking if users like the solution.

IDEO DT process (2024) includes six steps: (a) Frame a question, (b) Gather inspiration, (c) Generate ideas, (d) Make ideas tangible, (e) Test to learn, (f) Share the story. UK Design Council (2005) proposed the Double Diamond process model. This model includes defining and developing steps around two rhombuses in convergent and divergent thinking phases. While the first diamond is about understanding people and their needs, the second diamond includes processes related to the solutions produced. Although there is diversity in the naming of the steps in these models, the creative problem-solving processes are similar (Dam & Siang, 2021).

Significance of DBL in CT

Education systems considered DT and CT within the scope of 21st-century high-order thinking skills. Both ways of thinking focus on developing individuals' problem-solving skills, whether digital or not. Considering this goal, DT involves students in a process that requires them to use divergent thinking skills in understanding, defining, and solving human-oriented problems. CT skills also provide a framework for learners to find more generalized solutions by following processes such as understanding a problem, breaking it down into smaller parts, abstracting, and modeling. Both ways of thinking offer a series of steps for individuals to strengthen their competencies in tackling real-world problems. DBL offers an advantageous ground for learners as it involves exploring real problems using creative thinking and human-oriented approaches beyond a process in which innovative solutions are produced and logically solve problems. At the same time, creativity is considered one of the core components of CT (Korkmaz et al., 2017; Yildiz Durak, 2020, 2024).

Advantages of DBL for CT

The connection between DT and CT is essential because schools and universities increasingly integrate these thinking methods into their curriculum to address problems even though they are typically taught separately (Kelly & Gero, 2021). More studies are needed to integrate DBL with CT. Several studies in this context emphasize the positive role of DBL in fostering CT. For example, Saritepeci (2020) investigates the effect of DBL activities and programming education on CT skills in ninth-grade students, finding that both groups showed significant positive development in CT levels after a fourteen-week study, with DBL activities having a similar effect on CT skills compared to programming education. Jun et al. (2017) found that DBL using the Scratch Program and Creative Computing Guidebook from MIT improved CT in elementary school students more effectively than traditional teaching methods. Matere and colleagues (2023) investigated the application of DBL with primary school students to enhance CT and knowledge creation in creating end products. Their findings showed that while there was no noticeable distinction in the progress of CT between the test and control groups, there was a significant difference in the end product's design. These findings suggest that integrating both teaching modalities into elementary school curricula may have benefits. These results indicate the possible advantages of combining both instructional approaches in elementary school curriculums. A DBL technique was employed in a study to enhance the CT skills of primary school pupils. The results showed a considerable rise in the kids' self-perceived CT skills, with fluctuations in different learning phases. The finding highlights the significance of collaboration throughout the design phase of the learning process (Li et al., 2021).

Intersecting Processes of CT and DT

DT and CT include the problem-solving process. Originating in computer science, CT is a problem-solving methodology that may be used to tackle various personal and professional issues. It frequently involves using computers and provides a step-by-step representation of solutions for related problems (Kelly & Gero, 2021). DT similarly focuses on understanding the problem through empathy, developing ideas, and reaching the optimum solution based on testing and feedback. In this context, DT steps and CT components can be used together to solve real-life problems. Both ways of thinking require creativity and critical thinking. CT can be used to collect, analyze, and visualize data related to needs by following the empathy step processes during the problem's recognition and definition (Yildiz Durak, 2020; Yildiz-Durak et al., 2023a, 2023b). Additionally, DT and CT skills can be intersected in developing ideas for possible solutions and in the prototyping and testing phase. Figure 3.1 depicts how CT and DT intersect and can be used together.

However, some differences between CT and DT are also worth noting. The synthesis of DT and CT suggests that they produce solutions at opposite ends of a continuum, with design solutions being concrete to the problem and rarely transferable, while CT solutions generally apply to new issues (Kelly & Gero, 2021).

Pedagogically, integrating DBL and CT can help students become more holistic problem solvers. When both ways of thinking are used together, they can be used to recognize and define real-life problems, generate solutions, and test. In this context, CT components can be combined to develop more human-centered projects. Project-based learning can be used as a pedagogical basis for integrating DBL and CT.

Summary of Key Points

The essential components of DBL include developing an authentic learning environment, using a learner-centered approach, stimulating research and inquiry, supporting iteration, prototyping, and promoting deep learning. The essential components of DBL can potentially encourage a more immersive, student-centered approach to learning. DT and CT skills can be combined in the context of higher-order thinking

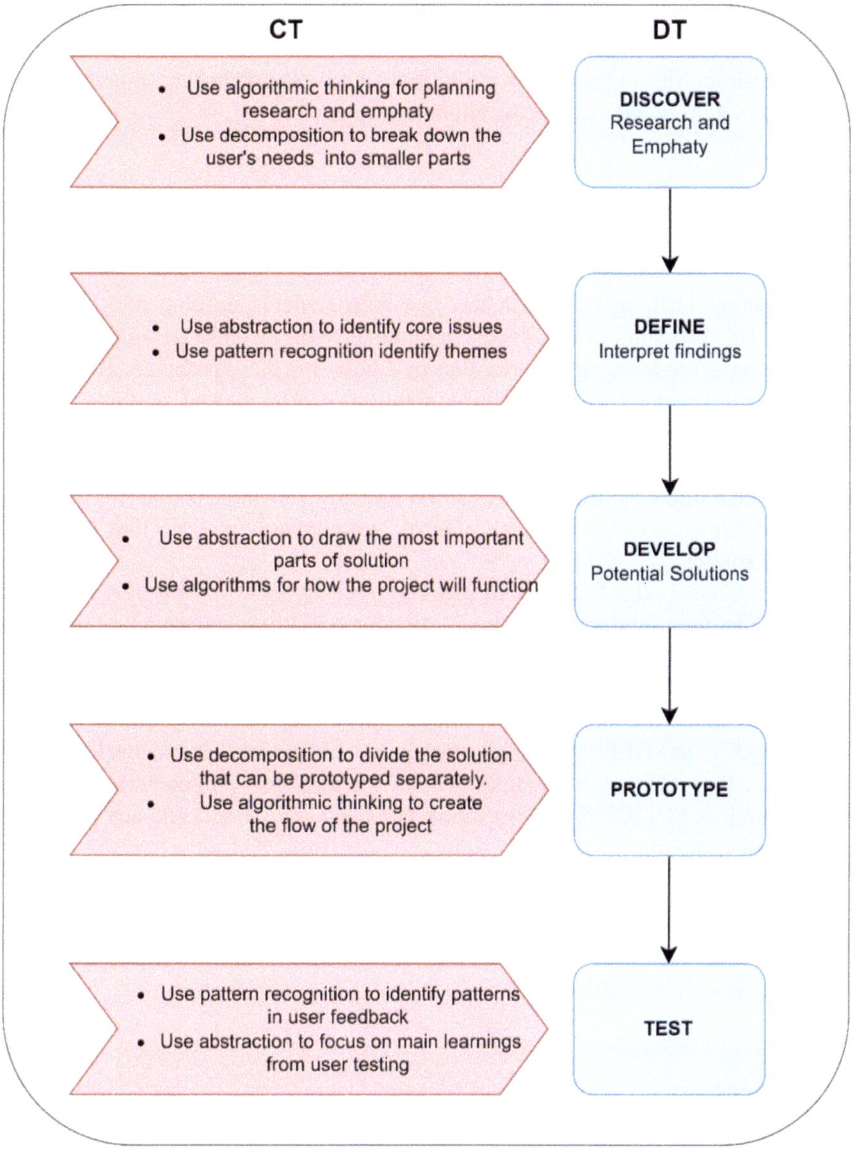

Fig. 3.1 Teaching tips for the intersection of CT and DT

skills of the twenty-first century. The integration of CT and DT supports students in becoming more holistic problem solvers while supporting them in recognizing real-life problems, generating solutions, and testing them. While DT requires participation in the process of solving human-centered problems with different thinking skills, CT provides a framework for finding general solutions by breaking problems into smaller pieces and through modeling. In conclusion, DBL has the potential to develop a new generation of holistic problem solvers with the abilities and mentality required to approach difficulties in the real world with creativity, empathy, and ingenuity as it continues to gain popularity in educational settings.

Implications for Future Research and Practice

By combining DBL and CT, teachers can help students develop into more all-encompassing problem solvers. Teachers should provide all children with more equitable and engaging learning opportunities by addressing these problems. Real-world problems can be identified and defined, solutions can be created, and tests can be conducted using this integration. CT framework could provide the basis for the generalizability of these solutions. The intersection suggested in the current study can be used in future studies. Further investigation into how this intersection can be established and its effects on students' learning and problem-solving skills may be crucial for future studies.

Future studies in this field should focus on figuring out how DBL affects student learning outcomes, how well it fosters 21st-century abilities, and what best practices are for putting DBL into practice in various educational contexts.

Teachers need assistance and training to apply DBL in the classroom. It is recommended that studies be conducted on professional development programs for teachers integrating CT and DBL teaching competencies. In this context, teachers' capabilities in teaching CT can be a starting point for teachers' proficiencies for DBL. Yildiz-Durak et al. (2023a, 2023b) found that teaching CT self-efficacy positively affects DBL teaching.

Also, to support students in improving their problem-solving skills through CT and DT and to ensure that they align with the DBL principles, schools and other educational institutions may need to review their curricula and assessment procedures.

References

Altay, B. (2014). User-centered design through learner-centered instruction. *Teaching in Higher Education, 19*(2), 138–155.

Azizan, S. A., & Abu Shamsi, N. (2022). Design-based learning as a pedagogical approach in an online learning environment for science undergraduate students. *Front. Educ., 7*, 860097.

Carella, G., Cautela, C., Melazzini, M., Pei, X., & Schmittinger, F. (2023). Design thinking for entrepreneurship: An explorative inquiry into its practical contributions. *The Design Journal, 26*(1), 7–31.

Carlgren, L., Elmqvist, M., & Rauth, I. (2014). Exploring the use of design thinking in large organizations: Towards a research agenda. *Swedish Design Research Journal, 11*, 55–63.

Cha, H. (2019). Applying design thinking to the educational problems: A student-centered instructional approach and practice in an undergraduate course. *Educational Technology International, 20*(1), 83–107.

Chandrasekaran, S., Stojcevski, A., Littlefair, G., & Joordens, M. (2013). Project-oriented design-based learning: Aligning students' views with industry needs. *International Journal of Engineering Education, 29*(5), 1109–1118.

Dam, R. F., & Siang, T. Y. (2020). Design thinking: Get started with prototyping. *Interaction Design Foundation.* https://library.parenthelp.eu/wp-content/uploads/2021/03/www.interaction-design.org_.pdf

Dam, R. F., & Siang, T. Y. (2021). *What is design thinking and why is it so popular?.* https://eclass.aueb.gr/modules/document/file.php/MISC418/Lecture3_Reading-Design_Thinking.pdf

Doppelt, Y., Mehalik, M. M., Schunn, C. D., Silk, E., & Krysinski, D. (2008). Engagement and achievements: A case study of design-based learning in a science context. *Journal of Technology Education, 19*(2), 22–39.

Ejsing-Duun, S., & Skovbjerg, H. M. (2019). Design as a mode of inquiry in design pedagogy and design thinking. *International Journal of Art & Design Education, 38*(2), 445–460.

Gómez Puente, S. M., van Eijck, M., & Jochems, W. (2011). Towards characterising design-based learning in engineering education: a review of the literature. *European Journal of Engineering Education, 36*(2), 137–149.

Gonera, A., & Pabst, R. (2019). The use of design thinking in transdisciplinary research and innovation consortia: Challenges, enablers, and benefits. *Journal of Innovation Management, 7*(3), 96–122.

Grau, S. L., & Rockett, T. (2022). Creating student-centred experiences: using design thinking to create student engagement. *The Journal of Entrepreneurship, 31*(2_suppl), S135–S159.

Hasso-Plattner Institute of Design (2016). *An introduction to design thinking PROCESS GUIDE.* https://web.stanford.edu/~mshanks/MichaelShanks/files/509554.pdf

Hehn, J., Uebernickel, F., Stoeckli, E., & Brenner, W. (2018, March). Designing human-centric information systems: towards an understanding of challenges in specifying requirements within design thinking projects. In *Multikonferenz Wirtschaftsinformatik* (pp. 1051–1062).

IDEO. (2024). *The design thinking process.* https://www.ideou.com/blogs/inspiration/design-thinking-process

Inder, S., & Withell, A. (2013). *A case study of authentic learning underpinned by design thinking and industry collaboration.* https://openrepository.aut.ac.nz/server/api/core/bitstreams/31e11e62-b209-44a7-89fe-06f8905c9e0d/content

Jun, S., Han, S., & Kim, S. (2017). Effect of design-based learning on improving computational thinking. *Behaviour & Information Technology, 36*(1), 43–53.

Kelly, N., & Gero, J. S. (2021). Design thinking and computational thinking: A dual process model for addressing design problems. *Design Science, 7*, e8.

Korkmaz, O., Cakir, R., & Ozden, M. Y. (2017). A validity and reliability study of the Computational Thinking Scales (CTS). *Computers in Human Behavior, 72*, 558–569.

Li, X., Zhang, Y., & Huang, J. (2021, September). Testing a design-based learning approach to enhance elementary students' computational thinking with experience-sampling method. In *Proceedings of the 3rd world symposium on software engineering* (pp. 17–22).

Matere, I. M., Weng, C., Astatke, M., Hsia, C. H., & Fan, C. G. (2023). Effect of design-based learning on elementary students computational thinking skills in visual programming maker course. *Interactive Learning Environments, 31*(6), 3633–3646.

Micheli, P., Wilner, S. J., Bhatti, S. H., Mura, M., & Beverland, M. B. (2019). Doing design thinking: Conceptual review, synthesis, and research agenda. *Journal of Product Innovation Management, 36*(2), 124–148.

Nawang, A. A. S. M. A. (2022, June). Digital marketing ın the era of socıety 5.0 by applying design thinking. *International Conference on Industrial Revolution 4.0, 1*(1), 284–295.

Pande, M., & Bharathi, S. V. (2020). Theoretical foundations of design thinking–A constructivism learning approach to design thinking. *Thinking Skills and Creativity, 36*, 100637.

Pereira, L., Parizi, R., Prestes, M., Marczak, S., & Conte, T. (2021, March). Towards an understanding of benefits and challenges in the use of design thinking in requirements engineering. In *Proceedings of the 36th annual ACM symposium on applied computing* (pp. 1338–1345).

Prayogi, S., Ardi, R. F. P., El Yazidi, R., Tseng, K. C., & Mustofa, H. A. (2023). The analysis of students' design thinking in inquiry-based learning in Routine university science courses. *International Journal of Essential Competencies in Education, 2*(1), 1–14.

Saritepeci, M. (2020). Developing computational thinking skills of high school students: Design-based learning activities and programming tasks. *The Asia-Pacific Education Researcher, 29*(1), 35–54.

Seitamaa-Hakkarainen, P. (2011). Design based learning in crafts education: Authentic problems and materialization of design thinking. *Design Learning and Well-Being, 4*, 3–14.

UK Design Council. (2005). *The Double Diamond A universally accepted depiction of the design process.* https://www.designcouncil.org.uk/our-resources/the-double-diamond/

Wagner, T. R. (2014). *Students' and Teachers' perceptions of the benefits and challenges of design-based learning in a middle school classroom* [Doctoral dissertation, Virginia Tech].

Weng, C., Chen, C., & Ai, X. (2023). A pedagogical study on promoting students' deep learning through design-based learning. *International Journal of Technology and Design Education, 33*(4), 1653–1674.

Yalçın, V., & Erden, Ş. (2021). The effect of STEM activities prepared according to the design thinking model on preschool children's creativity and problem-solving skills. *Thinking Skills and Creativity, 41*, 100864.

Yildiz Durak, H. (2020). The effects of using different tools in programming teaching of secondary school students on engagement, computational thinking and reflective thinking skills for problem solving. *Technology, Knowledge and Learning, 25*(1), 179–195.

Yildiz Durak, H. (2024). The model for middle school students' computational identity. *Education and Information Technologies, 29*, 4223–4251.

Yildiz Durak, H., Atman Uslu, N., Canbazoğlu Bilici, S., & Güler, B. (2023a). Examining the predictors of TPACK for integrated STEM: Science teaching self-efficacy, computational thinking, and design thinking. *Education and Information Technologies, 28*(7), 7927–7954.

Yildiz Durak, H., Saritepeci, M., & Durak, A. (2023b). Modeling of relationship of personal and affective variables with computational thinking and programming. *Technology, Knowledge and Learning, 28*, 165–184. https://doi.org/10.1007/s10758-021-09565-8

Zhang, J. L. (2020). The application of human comprehensive development theory and deep learning in innovation education in higher education. *Frontiers in Psychology, 11*, 1605.

Zhu, L., Shu, L., Tian, P., Sun, D., & Luo, M. (2024). Facilitating students' design thinking skills in science class: an exploratory study. *International Journal of Science Education*, 1–22. https://doi.org/10.1080/09500693.2024.2309658

Chapter 4
Integrating Computational Thinking via AI-Based Design-Based Learning Activities

Mustafa Saritepeci, Hatice Yildiz Durak, and Esra Kidiman

Abstract This chapter discusses integrating computational thinking (CT) as an essential skill for generating effective solutions to complex problems in different disciplines through a design-based learning (DBL) approach. Furthermore, this chapter also addresses the role of artificial intelligence (AI) in education and highlights its potential to enhance students' CT skills through AI-based design-based learning activities. The application of AI in various educational activities, such as using AI as a collaboration tool, providing personalized feedback, and developing AI-supported learning frameworks, are discussed. The chapter proposes a framework for designing AI-supported DBL activities based on the TPACK framework, including knowledge domains related to DBL, content, and AI tools. It proposes extending this framework to design AI-supported design-based learning activities that combine CT knowledge with host content. Based on this framework, it also presents a detailed seven-step learning activity plan for integrating AI-supported DBL into CT instruction.

Keywords Computational thinking · Design-based learning · Generative AI in Education · AI-based DBL

Introduction

Computational thinking (CT) has been included in curricula as a crucial skill for individuals to solve complex problems effectively (Saritepeci, 2020; Saritepeci & Durak, 2022; Yildiz Durak et al., 2023b). CT is a cognitive skillset that includes problem interpretation, data analysis, algorithm development, decomposition, abstraction,

M. Saritepeci (✉) · H. Y. Durak
Department of Instructional Technology, Necmettin Erbakan University, Konya, Turkey
e-mail: mustafasaritepeci@gmail.com

E. Kidiman
Ministry of National Education, Ankara, Turkey

M. Saritepeci and H. Yildiz Durak (eds.), *Integrating Computational Thinking Through Design-Based Learning*, https://doi.org/10.1007/978-981-96-0853-9_4

pattern recognition, data visualization, and modeling in various fields (Berikan & Özdemir, 2020; Shute et al., 2017; Yildiz Durak, 2020).

On the other hand, design-based learning (DBL) encourages creative design processes to overcome real-world challenges. Saritepeci (2020) argues that it is essential to design basic processes to understand the mechanism of computer technologies and to use these technologies in line with individuals' basic needs and purposes. This skill requires acquiring problem-solving abilities, including breaking down complex problems into smaller, more manageable parts, abstraction, generalization, and utilizing various digital technologies in daily life (Bubica & Boljat, 2018). Advances in computer technologies and the application of future technologies in all disciplines require the acquisition of problem-solving skills and the competence to test innovative digital solutions in real-world scenarios. Chen et al. (2017) highlight that training based on specific software and hardware will eventually become dysfunctional due to the dynamic structure of developing technologies. At this point, it is critical to create a mechanism that provides learners with design and CT skills that can be transferred to solving everyday or future problems. Design-based learning activities (DBLA) can significantly contribute to teaching and reflecting on CT skills (Saritepeci, 2020).

On the other hand, developing CT skills is also related to competence in using technological tools (Yildiz Durak, 2024; Yildiz Durak et al., 2023a). Developing these skills is essential to creating a context that fosters thinking, learning, and creating. DBLA and experiences with new technologies, such as artificial intelligence (AI), can be practical ways to develop CT skills.

AI has recently been used in many tasks in daily life and various environments (Yildiz Durak, 2023). AI technologies are also actively addressed in education (Yu et al., 2019). The rapid development of AI has facilitated its widespread application in education (Kong et al., 2023). AI is a vast field capable of performing tasks that typically require human intelligence. AI technologies have the potential to facilitate the development of subject knowledge and critical skills at various educational levels and can be integrated into different learning environments (Jia et al., 2022). Integrating AI into these learning activities is a significant opportunity to improve students' CT skills.

However, DBL can more effectively foster creative design processes with AI technologies to overcome real-world challenges. Moreover, AI technologies such as data analysis tools, pattern recognition algorithms, decision-making engines, and data visualization techniques can be constructed to provide scaffolding in DBL processes for students interested in design and problem-solving tasks. In this context, AI-supported design-based learning activities have a significant potential for teaching and developing CT skills.

This chapter addresses the integration of CT into DBLA using AI technologies. AI can offer a new approach to be used individually and collaboratively with students, allowing them to experience the create-to-learn process from a new perspective. This chapter discusses integrating AI-based design-based learning activities (AI-DBLA) into CT training, the design of these activities, and how they can improve students' CT skills. Consequently, it aims to contribute to developing a comprehensive learning environment that uses AI technologies to enhance students' CT skills in the DBL context.

Role of AI in Computational Thinking and Design-Based Learning

DBL is a cyclical process that offers the opportunity to use design processes and aims to produce a product at the end of this process. It allows students to apply their knowledge and skills, test their ideas, receive feedback, and restructure their products. This process encourages students to think differently and share their work with others (Han & Bhattacharya, 2001). The design process consists of several stages, including problem definition, needs identification, information gathering, solution development, solution selection, alternative solution generation, appropriate solution selection, prototype design, development, testing, evaluation, and revision (Doppelt & Schunn, 2008; Hira & Hynes, 2019). These phases follow a cyclical pattern that allows for reversibility. Considered as a whole, DBL is a learning model that supports the acquisition of basic skills, such as applying knowledge and skills in the field, collaborating in groups, and fostering creative thinking.

The primary purpose of teaching CT is for students to discover how the computer system works and to adapt computer strategies to real-life problems (Durak & Saritepeci, 2018; Yildiz Durak, 2020). In this context, the learning design provided by DBL will provide significant contributions and advantages to teaching CT. AI can give students more effective learning experiences by combining CT and DBL. AI technologies can support students in applying knowledge, generating creative solutions, and developing critical thinking skills. Integrating AI applications can help students acquire the skills to address complex design-based problems.

AI can help students develop their computational thinking skills by assisting them in solving algorithmic problems, recognizing patterns, and improving their logical thinking abilities with ease. Natural language processing and deep learning technologies serve as feedback tools. AI is a valuable resource in learning environments, providing structure and enabling better control of situations. In Design-Based Learning, artificial intelligence can guide students in problem-solving processes. The role of AI in this guidance can be to help students make design decisions, analyze design decisions, and develop their projects. AI technologies can also monitor student progress through learning analytics and big data, providing personalized feedback.

AI Technologies for Computational Thinking and Design-Based Learning

AI is a scientific field with its methods, approaches, and technologies (Xu et al., 2021). Using or benefiting from this field should also be acquired as a skill. Instead of individuals solving problems independently, using tools that have a facilitating role in problem-solving is also a crucial problem-solving skill. Also, Defining the problem and determining the solution methods, choosing the most optimal solution method, and putting it to work often require individuals to produce a product in a field. This process involves utilizing knowledge of design principles. AI technologies enable people to create products by taking on the burden of processes that require a lot of knowledge and experience, such as writing code without knowing coding or creating qualified visual designs without having competent visual design skills.

Tasks Performed by AI for Computational Thinking

CT 2.0 (Tedre et al., 2021) is the solution to the inadequacy of existing CT content in the K12 process caused by the inclusion of AI. Utilizing AI for CT teaching through classification, prediction, and generation (Van Brummelen et al., 2019), CT 2.0 is the cutting-edge approach to CT education. In the CT training process, it is crucial to comprehend the purpose of these forms of use and the problems they aim to solve.

Numerous studies in computational thinking (CT) concentrate on programming skills and their development (Li et al., 2023). AI can be utilized in developing and teaching CT-related skills at various stages of plugged or unplugged activities related to programming. For instance, AI support can be used to create an algorithmic solution. The rationale for using AI in an algorithmic solution process can be illustrated with an example. The pandemic has led to the development and implementation of various technological solutions. These include projects such as using QR codes to prevent the entry of high-risk individuals into social areas, utilizing face recognition to check for mask-wearing, and alerting individuals to incorrect mask usage. While some of these projects may seem simple, they are crucial in mitigating the spread of the virus.

Additionally, some of these projects can be coded by children with the assistance of AI technology. For instance, if the QR code belongs to a personal cell phone and the person is not flagged as risky in the database, they will be granted access without AI assistance. One method to regulate entry to a mask-mandatory area is to use a machine-learning model to determine if individuals are wearing masks. The algorithm will include a model trained with a dataset of thousands of masked and unmasked face photos at the decision-making stage. This model will be able to detect accurately whether people are wearing masks.

Coding and AI tools appropriate for various age levels can be utilized in CT projects. Block-based tools provide programming and AI support. MIT's machine learning for kids project allows young children to create or use pre-made models in their design processes within the Scratch environment. Every day, examples of AI integration into CT teaching are being sought. New learned models are continually added to the tools, and the possibility of producing solutions to various problems is emerging.

Tasks Performed by AI for Design-Based Learning

Educators increasingly use design thinking skills and methods to enhance student creativity and problem-solving abilities (Jlang & Pang, 2023). The learning-by-design framework divides the learning process into two cycles: designing/redesigning and investigating/exploring. During the first stage, students utilize their existing knowledge to design or redesign something. In the second stage, designers utilize their existing designs and additional knowledge to expand their understanding (Van Zadelhoff et al., 2021). The connection between these two stages is established through the concepts of 'need to do' and 'need to know' (Wang et al., 2022). GenAI tools have significant potential as collaboration tools during these stages. For instance, if there is a problem-solving process that involves selecting the most suitable solution from multiple options. When designing, optimization is critical. By choosing and implementing this approach, GenAI tools can lead to more accurate decisions and more efficient processes and products.

On the other hand, using AI tools in the production phase can minimize the student's cognitive load and effort. GenAI tools provide numerous opportunities for students to develop problem-solving designs across various subjects. For instance, students can create a video to increase awareness about a particular topic. GenAI can assist in content creation, organization, flow, and storytelling within a specific scenario during the video preparation. Additionally, students can utilize GenAI tools to create digital content with customizable prompts, including parameters such as scenario, setting, characters, appearance, and emotions for the video.

Using Artificial Intelligence in Design-Based Learning to Improve Computational Thinking Skills

This section discusses incorporating CT skills into the DBL process by providing suggestions for using AI in various problem situations and examples of the use of AI.

Novel Approach to AI-Supported Design-Based Learning

DBL is used as a backbone, while AI technologies represent technological components that are constantly evolving and being added. In the K12 education era, AI is a new field of study that is taking shape depending on the creativity of teachers, students, and researchers. DBL has a very fundamental life cycle. Any new technology can be integrated into this lifecycle. Nevertheless, GenAI technology also offers powerful support that can be anchored in this lifecycle. Because GenAI tools have the potential to change productive capabilities and creative processes, and as they are redesigned, they can impact many sectors of society (Epstein et al., 2023).

The primary benefit of GenAI tools and services is the potential to significantly improve learning and productivity for novice learners or those less experienced in any task by reducing internal and external cognitive load (i.e., Brynjolfsson et al., 2023). Research can be tiring and complex for novice learners with limited prior knowledge of identifying design problems and developing solution ideas in a DBL process. GenAI tools and services can assist novice learners in accessing relevant information more quickly and easily. GenAI can also facilitate the difficulties encountered in generating and implementing design solutions. In complex processes like DBL, more meaningful learning processes can be created when learner performance is shaped by interaction with GenAI. However, the effect of GenAI on the performance of expert learners and professionals in any subject is relatively minor. Nonetheless, GenAI still offers significant advantages in terms of efficiency and time for professional learners and expert workers.

Especially in the last few years, the significant popularity of large language models (LLMs) (i.e., ChatGPT, Claude, Gemini) and the increasing number and capability of GenAI tools in areas such as visualization, motion animation, and video creation integrated with these LLMs have increased the focus on the use of AI in learning environments.

It is possible to get GenAI support in most of the DBL process. With these tools, it is possible to use all kinds of language-independent resources while researching and many artistic, literary, and visual tools while creating a solution for a solution and presenting products in different languages. More importantly, GenAI supports collaborative work in the DBL process by facilitating peer collaboration and connecting students with various backgrounds, interests, and expertise (Chan & Lee, 2023). In addition, with GenAI, students can access a large pool of information to prepare experiments, reports, and presentations, simulate experiments, analyze data, compose, produce creative writing and visual arts products, and have the chance to produce in different languages. Many design tools have started to offer AI support. Thanks to AI support, students are not limited to their skills and knowledge while designing. Therefore, using GenAI effectively should be seen as an essential skill. It is vital to support the use of GenAI with structured plans to gain the ability to use it.

Integration of AI-Based Design-Based Learning Activities into Computational Thinking Instruction

To integrate AI-DBLA in CT teaching, we will focus on a sample in-class activity that can be done in a course focusing on teaching computer science. First, the lesson should be phased in to ensure the integration of AI and CT. The contents of the stages can be exemplified as follows:

During the preparation phase, the AI tools that can be used in the project should be presented to the students, and they should be enabled to use these tools through some examples. As students' experience using the tools increases, the duration of this phase in the activities may gradually decrease. However, it is essential to improve students' prompt writing skills. Many recommendations support users in interacting correctly with GenAI tools. For example, Subramonyam et al. (2023) have determined the following three steps for the user to get effectual output with LLM tools: (1) what the task should be, (2) how to instruct the LLM to do the task, and (3) what to expect for the LLM's output in meeting the goal (Subramonyam et al., 2023). While providing a guide on correct text writing when using GenAI, Schuckart emphasized the importance of considering these technologies' ongoing development and improvement (Schuckart, 2023).

Content close to the problem situation, including basic CT concepts and the use of CT, should be presented. For example, if there is content that includes smart cities for the problem situation, the concept of IoT can be explained. A video shows how smart cities improve quality of life. In this case, a relationship can be established with CT topics, and relevant CT contents can be determined.

The teacher has tasked the students with developing a solution to the question to identify the integrated context-driven design problem: "Smart cities improve people's quality of life in an environmentally friendly manner; they are cities supported by technology. What kind of projects would you suggest for stray animals when designing a smart city?" The teacher asks the students to list their answers to this question.

Then, the students diversify and improve their answers using an AI chat generator like writecream.com. Below are the sample answers provided by a chat generator from writecream.com.

Prompt:

"What kind of projects would you suggest for stray animals when designing a smart city?"

Responses:

1. Smart Feeding and Water Stations
2. Smart Identification and Tracking
3. Smart Animal Control and Rescue
4. Public Awareness and Engagement
5. Smart Sterilization and Vaccination.

In the next stage, students are expected to identify which answers are in a structure that an information processing unit can solve. Students are confronted with many solutions at this stage. One of the essential steps is not to choose the most appropriate solution by examining the solutions but to synthesize a solution from these solutions. After this step, students, individually or as a group, determine the proper solution and reflect on how it was realized. If, in the case of developing an application, some variables may need to be defined. For instance, if the solution involves "solar-powered stations providing timed feeding and water," the visual design of the station is necessary. The station's visual design was prepared with the support of the Generative AI application. For this project, visual design, electronic design, and coding stages may be sub-working groups.

The solution is realized by selecting a suitable coding tool for the algorithmic solution whose flowchart is created. In the sample application, the application named Pictoblox was preferred.

"Technologies used to analyze animal species and conditions, identify unhealthy or injured animals, and notify the relevant unit." could be another solution. In this case, the student uses machine learning (ML) models for the solution. For the application, the student first develops a flowchart. It is coded using a block-based coding tool, for example, Pictoblox, which is appropriate for the flowchart. The ML model recognizes the animal in the image taken from the camera and determines the health condition.

In the examples, we examined a CT application's design and development process supported by generative AI and ML. After this stage, students can test their designs with their peers and teachers, and the design can be improved by going through the design and development process again.

Framework for Designing AI-Supported Design-Based Learning Activities

Generative AI (GenAI) as an expert collaboration or a guidance tool can provide more efficient and effective DBL processes in many aspects. The development of GenAI technologies and their potential for educational processes and implementation has led to the idea of revising and extending the TPACK framework, which is frequently used in ensuring the effective use of technology in education (e.g., Celik, 2023; Mishra et al., 2023). Considering these extensions, we propose to replace pedagogical knowledge with DBL and technological knowledge with AI tool knowledge in the TPACK framework to design effective AI-DBLA. We present knowledge domains that can be mapped to three knowledge domains based on the TPACK framework for designing AI-DBLA (Fig. 4.1): Design-based learning knowledge (DBLK), content knowledge (CK), and AI tool knowledge (AIK).

DBLK is the knowledge of DBL methods and techniques in teaching and learning processes and how to apply them in the context of instructional goals and objectives.

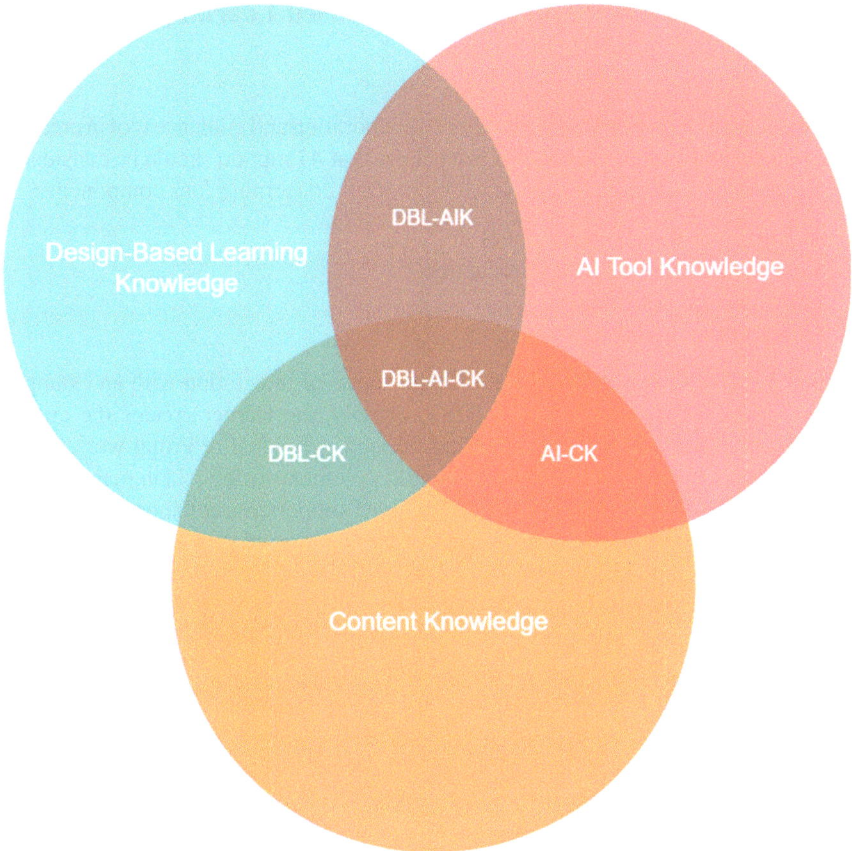

Fig. 4.1 DBL-AI-CK framework

AIK is the knowledge of interacting with AI tools and using their essential functions (Celik, 2023). A deep understanding of the capabilities and limitations of each tool is necessary for designing an effective AI-DBLA. CK is the knowledge of basic facts, concepts, theories, and procedures related to the subject, domain, or discipline to be taught (Schmid et al., 2020). DBL-CK is the knowledge of creating and operating DBL processes appropriate to the content taught in a particular discipline. DBL-AIK is the knowledge of how to optimize DBLA processes using AI tools. AI-CK involves.

- understanding the interrelationship between AI tools and services and instructional content,
- identifying the most appropriate AI tools to convey this content and
- using them appropriately.

DBL-AI-CK is the knowledge of effectively using the most appropriate AI tools and services for learners to acquire meaningful learning through DBLA activities.

How to Design AI-Supported Design-Based Learning Activities

In DBLA, we emphasize using GenAI as a collaboration and guidance tool in creating an AI-supported structure. However, we believe that AI support should be considered for the learning environment and teaching process concerning four components:

- AI support as a collaboration tool
- AI support for cognitive load reduction
- AI support for scaffolding
- AI support as a guidance tool.

It is possible to concretize these different forms of integration with an example. In an AI-based digital storytelling (DST) activity, the learner creates the critical elements of a story concerning a context through individual or group work. Then, they can develop stories that reference these key elements using AI language models (i.e., Saritepeci & Yildiz Durak, 2024a). Thus, AI support as a collaboration tool can enable students to create creative designs for authentic problems (Guan et al., 2020). In addition, it is emphasized in various studies that students who experience activities such as DST for the first time have difficulties in issues such as story writing, and this has the potential to negatively affect students' attitudes and motivation (e.g., Özüdoğru, 2021; Özüdoğru & Çakır, 2020; Sarıtepeci, 2016; Saritepeci & Çakır, 2019). Using AI as a collaboration tool can facilitate students' adaptation to such activities.

On the other hand, using AI tools in activities other than collaboration can reduce cognitive load. For instance, in a DST process, students must be familiar with various digital environments to transfer their storyboards from paper to digital formats. To become proficient in producing digital content using GenAI tools, learners must learn the details about each tool, which can cause a significant cognitive load. However, since prompts guide the digital content production processes in GenAI tools, it is often sufficient for the learner to have knowledge and application competence about the prompt frameworks that can be used for any purpose. In this way, the learner can use and produce AI tools designed for different purposes without significant usage experience and training.

DBL activities are time-consuming and require learners to exert effort outside class hours (Konaş & Durak, 2023; Saritepeci & Çakır, 2019; Saritepeci, 2021; Yildiz Durak, 2018). Therefore, it may not always be possible to access the support of an instructor. In such cases, AI tools can potentially support and guide students in and out of the classroom. The student's access to an AI course assistant, which he/she can reach when he/she needs any help or wants to get feedback on the design he/she has prepared, allows the student to progress independently and at his/her own pace. As a matter of fact, Sarıtepeci and Yildiz Durak (2024b) used a ChatGPT-based ChatBot as a course assistant to provide guidance and support to students in a paper game prototype project as an example of DBLA in the introduction to algorithms and programming course that they conducted online in the spring semester of 2022–2023. The students used the ChatBot as a support and guidance tool in determining the game's dynamics (i.e., purpose, mechanics, defining actions, characters, scoring system, game completion), developing an algorithm for the game, and identifying difficulties and errors encountered during this development process. Such ChatBots do not require any technical knowledge. With various applications and tools (e.g., Poe), ChatBots can be created for different purposes using prompts and training files.

Expanding the DBL-AI-CK Framework with CT as a Second CK

We propose to extend the DBL-AI-CK framework we present in this chapter to include CT in different content areas through AI-supported DBLAs. There are framework proposals in the literature (i.e., Lane et al., 2023; Zha et al., 2021) that include CT as a second CK in the TPACK framework for teaching CT through different disciplines. Based on this practical evidence in the literature, adding CT as a second CK domain to the DBL-AI-CK framework has significant potential for integration into AI-supported DBL teaching activities (Fig. 4.2). Such an integration involves focusing the teaching–learning process for discipline or subject area independent CT teaching on teaching both CT and host content knowledge (Waterman et al., 2020; Zha et al., 2021). Zha and colleagues argue that CT content knowledge shares a common content domain with similar core content domains such as Math. We take this one step further and argue that realizing this integration for any discipline is possible. Therefore, in the model, designing AI-DBLA that integrates CT knowledge (e.g., algorithm design, decomposition, abstraction, pattern recognition, generalization) related to the topic to be conveyed in the host content will provide meaningful learning experiences in the related discipline and CT teaching. The outcomes of interventions involving CT integration with host contents, such as writing education (Wu et al., 2024) and music education (i.e., Bell & Bell), support this argument. In their study, Wu et al. reported that a CT-based teaching intervention in writing education significantly improved students' writing and CT-related skills (algorithmic thinking, decomposition, isolation, generalization, generalization, and evaluation).

Figure 4.2 shows how CT is integrated as a core component in designing and implementing AI-DBLA. This framework extends CT to intersect with host content, allowing educators and instructional designers to use AI technologies to facilitate meaningful learning experiences and develop CT skills across different content areas.

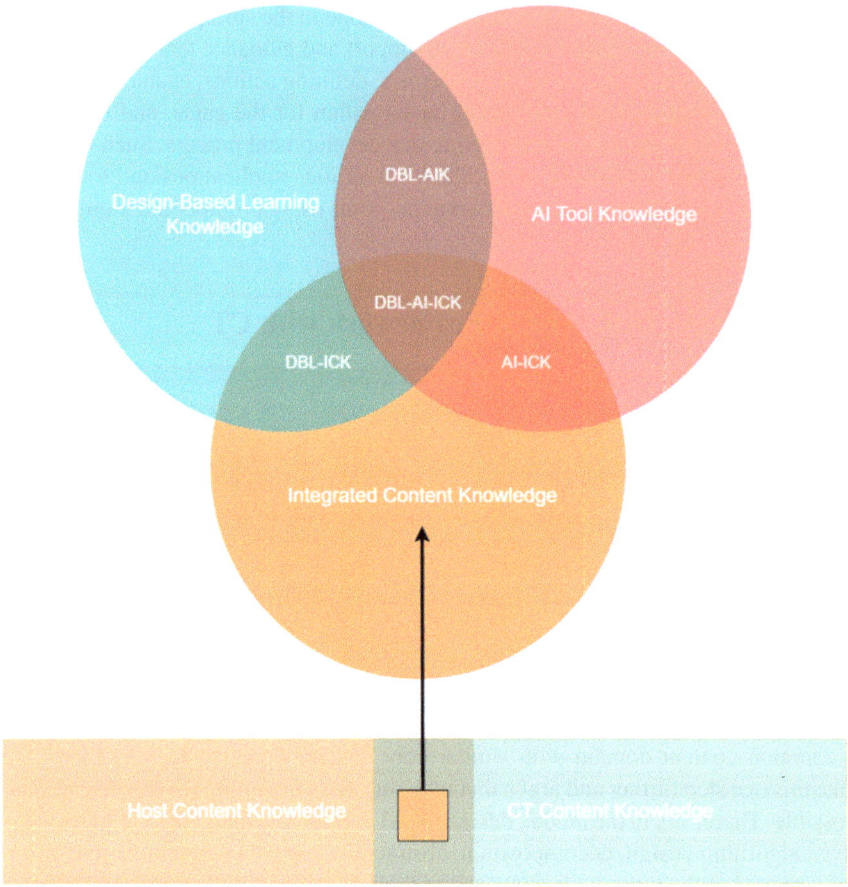

Fig. 4.2 CT integrated DBL-AI-CK

Activity Plan for AI-Supported Design-Based Learning for Developing Computational Thinking Skills

Taking into account the knowledge areas offered by the CT-integrated DBL-AI-CK extended framework for trainers to design effective learning processes, we propose a seven-step learning activity plan: (i) Preparation, (ii) Determining CT and host context concepts, (iii) Identifying the integrated context-driven design problem, (iv) Improving the design idea, (v) AI-supported design implementation, (vi) Design optimization, (vii) Presentation and assessment.@@

(i) Preparation

- Before the event, conducting a pre-test of the AI tools is essential to ensure their suitability.

- Introduce the AI tools that will be used and provide sufficient training and guidance on their utilization.
- Additionally, online documentation such as a user manual or usage video link should be provided to assist with using the AI tools.

(ii) Determining CT and host context concepts

- Combine the related CT and host context components to identify the instructional content.
- Present relevant information, such as situations or events, that will enable the identification of design problems related to the integrated context.

(iii) Identifying the integrated context-driven design problem

- Get learners to reflect on the integrated context and propose ideas for potential design issues.
- Provide feedback on the design problems identified by the learners as a result of group or individual work.
- Ask learners to identify the limitations of the design problems and the possibilities for solutions.

(iv) Improving the design idea

- Enable learners to research the integrated context using AI tools.
- Encourage them to generate ideas for the design problem using techniques such as brainstorming.
- Have them use AI tools to refine possible design ideas.
- Ask learners to share their design ideas and possible solutions.

(v) AI-supported design implementation

- Get learners to visualize or prototype their design ideas for solving the problem.
- Enable learners to improve their designs using appropriate AI tools.

(vi) Design optimization

- Encourage learners from AI tools to consider feedback and suggestions on their designs.
- Enable them to test the functionality of their designs when solving the problem.

(vii) Presentation and assessment

- Have learners present their designs about the problem situation they have identified.
- Have learners review each other's designs.
- Get feedback on the learners' experiences and use it to improve the activity.

Conclusions

Summary of Key Points

Integrating CT into DBLAs through AI support offers a remarkable perspective for fostering meaningful learning experiences in different content areas. In this chapter, integrating CT skills into different disciplines through DBL, one of the constructivist learning approaches, is discussed from the perspective of AI technologies. The potential of AI-based DBLA for teaching or developing CT skills is emphasized.

The chapter presents a framework for designing AI-supported DBLA and proposes a detailed seven-stage learning activity plan based on this framework. Considering this framework, the proposed seven-step learning activity plan provides an approach for educators to design effective learning processes that integrate CT and host content concepts. The plan includes the basic steps of preparation, identifying the integrated context-driven design problem, and implementing AI-assisted design.

Implications for Future Practice and Research

The CT-integrated DBL-AI-CK framework and the learning activity plan we present, considering this framework, encourage the use of AI-supported DBLA in integrating CT in different disciplines and contexts. Including CT in the DBL-AI-CK framework has the potential to impact the teaching and learning process significantly. In practice, this framework and the teaching activity plan can contribute to developing students' creative problem-solving and innovative product development skills. Future research could address the design of professional development programs for teachers to create effective CT integration skills based on the DBL-AI-CK framework. Students can develop their CT skills through these activities while engaging in real-world problem-solving scenarios. In addition, interventions can be developed in which learning activities based on the presented seven-step activity plan are employed, and their impact on students' domain and CT knowledge development is addressed.

References

Berikan, B., & Özdemir, S. (2020). Investigating "problem-solving with datasets" as an implementation of computational thinking: A literature review. *Journal of Educational Computing Research, 58*(2), 502–534.

Brynjolfsson, E., Li, D., & Raymond, L. R. (2023). *Generative AI at work* (No. w31161). National Bureau of Economic Research. https://www.nber.org/system/files/working_papers/w31161/w31161.pdf

Bubica, N., & Boljat, I. (2018). Assessment of computational thinking. In *Proceedings of the conference on computational thinking education* (CTE 2018). Retrieved from https://www.croris.hr/crosbi/publikacija/resolve/irb/958252

Celik, I. (2023). Exploring the determinants of artificial intelligence (Ai) literacy: Digital divide, computational thinking, cognitive absorption. *Telematics and Informatics, 83*, 102026.

Chan, C. K. Y., & Lee, K. K. (2023). The AI generation gap: Are Gen Z students more interested in adopting generative AI such as ChatGPT in teaching and learning than their Gen X and millennial generation teachers? *Smart Learning Environments, 10*(1), 60.

Chen, G., Shen, J., Barth-Cohen, L., Jiang, S., Huang, X., & Eltoukhy, M. (2017). Assessing elementary students' computational thinking in everyday reasoning and robotics programming. *Computers & Education, 109*, 162–175.

Doppelt, Y., & Schunn, C. D. (2008). Identifying students' perceptions of the important classroom features affecting learning aspects of a design-based learning environment. *Learning Environments Research, 11*, 195–209.

Durak, H. Y., & Saritepeci, M. (2018). Analysis of the relation between computational thinking skills and various variables with the structural equation model. *Computers & Education, 116*, 191–202.

Epstein, Z., Hertzmann, A., Investigators of Human Creativity, Akten, M., Farid, H., Fjeld, J., Frank, M.R., Groh, M., Herman, L., Leach, N., Mahari, R., & Smith, A. (2023). Art and the science of generative AI. *Science, 380*(6650), 1110–1111.

Guan, C., Mou, J., & Jiang, Z. (2020). Artificial intelligence innovation in education: A twenty-year data-driven historical analysis. *International Journal of Innovation Studies, 4*(4), 134-147. https://doi.org/10.1016/j.ijis.2020.09.003

Han, S., & Bhattacharya, K. (2001). Constructionism, learning by design, and project-based learning. In M. Orey (Ed.) *Emerging Perspectives on Learning, Teaching, and Technology*. Retrieved from https://pirun.ku.ac.th/~btun/papert/design.pdf

Hira, A., & Hynes, M. M. (2019). Design-based research to broaden participation in pre-college engineering: Research and practice of an interest-based engineering challenges framework. *European Journal of Engineering Education, 44*(1–2), 103–122.

Jia, F., Sun, D., Ma, Q., & Looi, C. K. (2022). Developing an AI-Based learning system for L2 learners' authentic and ubiquitous learning in English language. *Sustainability, 14*(23), 15527.

Jlang, T., & Pang, R. (2023). Educators increasingly use design thinking skills and methods to enhance student creativity and problem-solving abilities. *Geotechnics, 3*(4). https://doi.org/10.3390/geotechnics3040055

Konaş, Z. Ş, & Durak, H. Y. (2023). Examination of the effects of digital story applications on digital parenting and technology usage attitudes. *Instructional Technology and Lifelong Learning, 4*(1), 15–47.

Kong, S. C., Cheung, W. M. Y., & Zhang, G. (2023). Evaluating an artificial intelligence literacy programme for developing university students' conceptual understanding, literacy, empowerment and ethical awareness. *Educational Technology & Society, 26*(1), 16–30.

Lane, W. B., Galanti, T. M., & Rozas, X. L. (2023). Teacher Re-novicing on the path to integrating computational thinking in high school physics instruction. *Journal for STEM Education Research*, 1–24.

Li, X., Xie, K., Vongkulluksn, V., Stein, D., & Zhang, Y. (2023). Developing and testing a design-based learning approach to enhance elementary students' self-perceived computational thinking. *Journal of Research on Technology in Education, 55*(2), 344–368.

Mishra, P., Warr, M., & Islam, R. (2023). TPACK in the age of ChatGPT and Generative AI. *Journal of Digital Learning in Teacher Education, 39*(4), 235-251. https://doi.org/10.1080/21532974.2023.2247480

Özüdoğru, G. (2021). Digital storytelling in education from teachers' perspectives. *Bartın University Journal of Faculty of Education, 10*(2), 445–454.

Özüdoğru, G., & Çakır, H. (2020). An investigation of the effects of digital storytelling activities on engagement and writing self-efficacy. *Acta Didactica Napocensia, 13*(1), 105–116.

Sarıtepeci, M. (2016). *Investigation of the effectiveness of digital storytelling method in social studies course* [Unpublished PhD thesis, Gazi University, Institute of Education Sciences, Ankara, Turkey].

Saritepeci, M. (2020). Developing computational thinking skills of high school students: Design-based learning activities and programming tasks. *The Asia-Pacific Education Researcher, 29*(1), 35–54.

Saritepeci, M. (2021). Students' and parents' opinions on the use of digital storytelling in science education. *Technology, Knowledge and Learning, 26*(1), 193–213.

Saritepeci, M., & Çakır, H. (2019). The effect of digital storytelling activities used in a social studies course on student engagement and motivation. In *Educational technology and the new world of persistent learning* (pp. 24–55). IGI Global.

Saritepeci, M., & Durak, A. (2022). Adaptation of T-STEM CT scale to Turkish: Teacher self-efficacy and outcome expectancy for teaching computational thinking. *Research on Education and Psychology, 6*(Special Issue), 47–56.

Saritepeci, M., & Yildiz Durak, H. (2024a). Effectiveness of artificial intelligence integration in design-based learning on design thinking mindset, creative and reflective thinking skills: An experimental study. *Education and Information Technologies*, 1–35. https://doi.org/10.1007/s10639-024-12829-2

Saritepeci, M., & Yildiz Durak, H. (2024b). Redesigning introduction to algorithm and programming course under conditions of unanticipated online training: Empowered by GenAI. In *Proceedings of 4th international artificial intelligence and data science congress—ICADA* (pp. 21–25).

Schuckart, A. (2023). Introduction to work with GenAI. In *Proceedings of the 28th European conference on pattern languages of programs* (pp. 1–16).

Schmid, M., Brianza, E., & Petko, D. (2020). Developing a short assessment instrument for Technological Pedagogical Content Knowledge (TPACK. xs) and comparing the factor structure of an integrative and a transformative model. *Computers & Education, 157*, 103967. https://doi.org/10.1016/j.compedu.2020.103967

Shute, V. J., Sun, C., & Asbell-Clarke, J. (2017). Demystifying computational thinking. *Educational Research Review, 22*, 142–158.

Subramonyam, H., Pondoc, C. L., Seifert, C., Agrawala, M., & Pea, R. (2023). Bridging the Gulf of envisioning: Cognitive design challenges in LLM interfaces. arXiv preprint arXiv:2309.14459

Tedre, M., Denning, P., & Toivonen, T. (2021). CT 2.0. In *ACM international conference proceeding series*. https://doi.org/10.1145/3488042.3488053

Van Brummelen, J., Shen, J. H., & Patton, E. W. (2019, June). The popstar, the poet, and the Grinch: Relating artificial intelligence to the computational thinking framework with block-based coding. In *Proceedings of international conference on computational thinking education* (Vol. 3, pp. 160–161).

Wang, D., Luo, L., Luo, J., Lin, S., & Ren, G. (2022). Developing computational thinking: design-based learning and interdisciplinary activity design. *Applied Sciences, 12*(21), 11033.

Waterman, K. P., Goldsmith, L., & Pasquale, M. (2020). Integrating computational thinking into elementary science curriculum: An examination of activities that support students' computational thinking in the service of disciplinary learning. *Journal of Science Education and Technology, 29*, 53–64.

Wu, T. T., Silitonga, L. M., & Murti, A. T. (2024). Enhancing English writing and higher-order thinking skills through computational thinking. *Computers & Education, 213*(105012), 1–19. https://doi.org/10.1016/j.compedu.2024.105012

Yildiz Durak, H. (2018). Digital story design activities used for teaching programming effect on learning of programming concepts, programming self-efficacy, and participation and analysis of student experiences. *Journal of Computer Assisted Learning, 34*(6), 740–752.

Yildiz Durak, H. (2020). The effects of using different tools in programming teaching of secondary school students on engagement, computational thinking and reflective thinking skills for problem solving. *Technology, Knowledge and Learning, 25*, 179–195.

Yildiz Durak, H. (2023). Conversational agent-based guidance: Examining the effect of chatbot usage frequency and satisfaction on visual design self-efficacy, engagement, satisfaction, and learner autonomy. *Education and Information Technologies, 28*(1), 471–488.
Yildiz Durak, H. (2024). The model for middle school students' computational identity. *Education and Information Technologies, 29*, 4223–4251.
Yildiz Durak, H., Atman Uslu, N., Canbazoğlu Bilici, S., & Güler, B. (2023a). Examining the predictors of TPACK for integrated STEM: Science teaching self-efficacy, computational thinking, and design thinking. *Education and Information Technologies., 28*, 7927–7954. https://doi.org/10.1007/s10639-022-11505-7
Yildiz Durak, H., Saritepeci, M., & Durak, A. (2023b). Modeling of relationship of personal and affective variables with computational thinking and programming. *Technology, Knowledge and Learning, 28*(1), 165–184.
Yu, W. J., Jang, J. H., Ahn, J. M., Park, D. R., Yoo, I. H., Bae, Y. K., & Kim, W. Y. (2019). Design of artificial intelligence education program based on design-based research. *International Journal of Advanced Smart Convergence, 8*(4), 113–120.
Zha, S., Morrow, D. A., Curtis, J., & Mitchell, S. (2021). Learning culture and computational thinking in a Spanish course: A development model. *Journal of Educational Computing Research, 59*(5), 844–869.
Xu, Y., Liu, X., Cao, X., Huang, C., Liu, E., Qian, S., & Liu, X. (2021). Artificial intelligence: A powerful paradigm for scientific research. *The Innovation, 2*(3), 100150. https://doi.org/10.1016/j.xinn.2021.100179
Van Zadelhoff, B., Rahimi, E., & Barendsen, E. (2021). Principles to facilitate design-based learning environments for programming in secondary education while making learning visible in an authentic way. In *Proceedings of the 21st Koli Calling International Conference on Computing Education Research* (pp. 1-10). https://doi.org/10.1145/3488042.3488067

Chapter 5
Computational Thinking Integration by Design-Based Learning Perspective Implementation Example of AI

Şeyhmus Aydoğdu

Abstract The use of artificial intelligence (AI) tools is becoming increasingly widespread in educational sciences. In addition to the use of these tools as support in the educational context, it is very important to gain knowledge and skills for the design and development of tools to ensure reliability and effectiveness in the process of using these technologies by pre-service teachers. Although there are many studies on teaching machine learning (ML) processes, which is an application area of AI, to young age groups, there is a limited number of studies on teaching these processes to pre-service teachers. When the realization steps of ML are examined, it is seen that design-based learning (DBL) activities are suitable for teaching these steps to pre-service teachers. In this chapter, an implementation example of AI activities prepared based on DBL perspective for computational thinking (CT) skills was realized. This chapter is organized as follows: developing pre-service teachers' CT skills in the first section, teaching AI to pre-service teachers in the second section, CT integrated AI-Based activity design with DBL perspective in the third section, methodology of the implementation example in the fourth section, findings and discussion in the fifth section, recommendations for practitioners and researchers in the sixth section, AI-Based activity examples for CT Skills in the seventh section, and limitations and future researches in the eighth section.

Keywords Artificial intelligence · Design-based learning · Machine learning · Orange data mining

Ş. Aydoğdu (✉)
Department of Computer Education and Instructional Technologies, Nevşehir Hacı Bektaş Veli University, Nevşehir, Türkiye
e-mail: saydogdu@nevsehir.edu.tr

M. Saritepeci and H. Yildiz Durak (eds.), *Integrating Computational Thinking Through Design-Based Learning*, https://doi.org/10.1007/978-981-96-0853-9_5

Introduction

The development of computational thinking skills in pre-service teachers is very important in terms of enabling them to identify the steps towards a problem, carry out these steps systematically, and evaluate the solution. There are various studies (Aydoğdu, 2020; Gabriele et al., 2019; Umutlu, 2022; Villalustre & Cueli, 2023; Voon et al., 2023) on the development of pre-service teachers' computational thinking skills. Gabriele et al. (2019), in a study conducted with 141 pre-service teachers within the scope of a programming course, showed that most of the participants combined both design and programming skills in their school internships and reached medium–high level IT skills. Aydoğdu (2020) examined students' computational thinking skills and self-efficacy perceptions towards programming with block-based coding activities in the information technologies course. Similarly, Umutlu (2022) examined the computational thinking and programming skills of pre-service teachers in block-based coding activities in educational technologies course.

Research on the development of computational thinking skills is usually carried out through activities for teaching programming (Hsu et al., 2018). Although computational thinking is often equated with using computer technologies (Yadav et al., 2017), this skill refers to a thinking process that involves solving problems, designing systems, and understanding human behavior based on computer science concepts (Wing, 2006). It can be said that teachers' conceptualization of computational thinking is in line with the literature and this conceptualization does not differ according to teachers' disciplines (Sands et al., 2018). However, considering the integration of computational thinking skills into the education curriculum, it is of great importance to develop teachers' computational thinking skills beyond conceptualization (Rajapakse Mohottige et al., 2024). Dong et al. (2024) examined 38 studies on the development of computational thinking skills of pre-service teachers and grouped the studies on the development of computational thinking skills of pre-service teachers under the categories of CT courses, programming applications, educational robot training programs, CT module, CT projects and CT seminars. The fact that only 5 of the studies examined in the re-search have CT module as the training method shows that the effectiveness of the modules to be created by integrating computational thinking skills into different contexts should be investigated.

Teaching AI to Pre-service Teachers

Artificial intelligence is a field of study that produces computers, machines, and other artifacts with human-like intelligence and the resulting innovations and developments (Chen et al., 2020). Machine learning, a sub-field of artificial intelligence, deals with computers making inferences from data on specific topics using algorithms and statistical methods (Alpaydin, 2021). Although most of the applications used in the context of educational sciences today are directly called artificial intelligence-based, these applications are generally developed depending on the machine learning field of study.

Teachers can use AI-supported tools for generating content, assessing students, and developing their knowledge, skills, and competencies (Sarkar & Ahammed, 2023). AI tools have been actively used in fields of study such as adaptive learning environments (Gligorea et al., 2023; Kabudi et al., 2021), recommendation systems (Zhang et al., 2021), classroom management (Wang, 2021), automated scoring (Ramesh & Sanampudi, 2022). However, although artificial intelligence tools are considered effective in different contexts, it should not be forgotten that these tools create models based on data and make inferences according to these models. This means that, unlike algorithm-based systems, they do not always produce a single correct answer, but rather variable and potentially incorrect answers over time. In addition, the low-reliability levels of the results produced using these tools in some fields (e.g. automated essay scoring) (Ferrara & Qunbar, 2022) indicate that the results to be produced from these tools cannot be used directly yet.

In addition to the use of artificial intelligence and machine learning applications as supportive tools in learning environments, it is very important to provide individuals with the steps followed in the process of creating these applications and the knowledge and skills for creating applications to ensure the reliability and effectiveness of the applications. There are different studies on teaching machine learning from young children to adults (Kaspersen et al., 2021; Ojo, 2024; Sanusi, 2021; Sanusi et al., 2022, 2023; Shamir & Levin, 2022; Sundberg & Holmström, 2024; Tedre et al., 2021; Vartiainen et al., 2020). When these studies are analyzed, it is seen that the majority of the studies include artificial intelligence teaching activities for K-12 level students. Ojo (2024) conducted a study on the development of pre-service teachers to promote the teaching of artificial intelligence at the K-12 level. In this study, a professional development program was prepared to increase pre-service computer science teachers' knowledge of artificial intelligence. Artificial intelligence tools are widely used in daily life and include opportunities for use in learning processes. The limited number of studies on pre-service teachers within the scope of artificial intelligence teaching shows the necessity of preparing curriculums specific to the field of teaching and examining the effect of these curriculums on pre-service teachers.

CT Integrated AI-Based Activity Design with DBL Perspective

To prepare an AI-based activity design in which computational thinking skills are integrated, it is first necessary to examine computational thinking skills and machine learning application steps. There are differences in the literature regarding the steps of computational thinking processes. In Selby and Woollard's (2014) study, the consensus skills are thought process, abstraction, decomposition, algorithmic design, evaluation, generalization, and automation. Similarly, Dong et al. (2024)

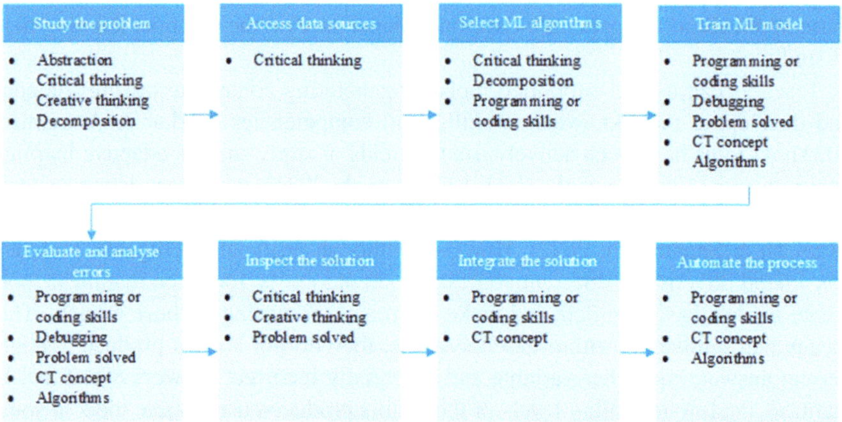

Fig. 5.1 Linking machine learning implementation steps and pre-service teachers' computational thinking skills

classified pre-service teachers' computational thinking skills as abstraction, critical thinking, decomposition, programming or coding skills, debugging, CT concepts, algorithms, creative thinking, and problem-solving. The relationship between the machine learning implementation steps for an AI-based activity design (Géron, 2022) and pre-service teachers' computational thinking skills is shown in Fig. 5.1.

How to connect with computational thinking skills as part of the machine learning application steps is explained below:

- Study the problem: Artificial intelligence or machine learning applications are designed to create a model by discovering the pattern in the data for problems for which algorithms cannot be written and to make predictions according to the model created. Therefore, the ability to write an algorithm for a problem should be examined first. The input and output variables of the problem should be clearly defined before the solution methods or data sources are determined. This step is directly related to abstraction, critical thinking, creative thinking, and decomposition skills. In this step, it is necessary to determine the main idea and search for information related to it, create a logical fiction between input and output variables, and determine the solution methods by decomposing the problem into sub-problems, if any.
- Accessing data sources: One of the most important steps in machine learning is accessing data sources. When teaching machine learning in the classroom, the teacher needs to provide students with information about the data sources that can be used. Accessing data sources is related to critical thinking skills. In this step, critical thinking skills are required for the practitioner to make associations between inputs and outputs in the accessed data sources and to integrate data from different data sources.
- Selecting ML Algorithms: The choice of ML algorithm is determined by the objective of the problem and the data structure to be used. In this step, algorithms

are selected according to the preferred ML approach and the type of variables in the data set. This step involves critical thinking, decomposition, programming, or coding skills. Critical thinking and decomposition skills are required to determine the appropriate algorithms and input and output variables for the data set, and programming or coding skills are required to determine the packages or libraries to be used for the selected algorithms.
- Train ML Model: In this step, the process of creating a model using the data set with the determined machine learning algorithms is carried out. This step requires programming or coding skills to define the data sets for the algorithms and modeling, debugging skills to perform the debugging steps in the developed solution, problem-solving, and CT concept skills to determine which parameters the models will be tested with, and algorithm skills to systematically design all steps in the training process.
- Evaluate and analyze errors: This step compares the generated models and determines which model to use. During model selection, errors are analyzed using metrics such as accuracy score. This step is related to programming or coding skills, debugging, problem-solving, CT concepts, and algorithms skills. Programming or coding skills are needed to compare the accuracy rates of the models and analyze the errors, debugging skills are needed to debug the errors made during the development process, and problem-solving and CT concept skills are needed to decide which algorithm is preferred according to which metric, and algorithms skills are needed to record the model results and make comparisons.
- Inspection of the solution: In this step, the predictions or calculations obtained from the models are examined. Critical thinking is required to determine the appropriateness of these calculations for the solution, creative thinking is required to identify different methods or approaches that can be used in the solution and problem-solving skills are required to determine if the solution is truly appropriate.
- Integrate the solution: In this step, the generated solutions are integrated into the system. This step requires programming and CT concept skills to put the solutions generated in the background to practical use.
- Automate the process: In this step, the above steps are repeated so that the machines perform a continuous modeling and learning process. This step requires programming or coding, CT concept, and algorithms skills.

Design-based learning environments are based on students' open exploration, trial-and-error learning, reflection, teamwork, and design creation using supportive tools in the context of project or problem-based learning environments (Zhang et al., 2018, 2020). In these environments, students use their theoretical knowledge to find answers to real-world problems (Azizan & Abu Shamsi, 2022). Considering the principles of project-based learning (Condliffe, 2017) and problem-based learning processes (Nilson, 2016), the strategies to be followed in teaching machine learning implementation steps are listed below:

- Associate with previous knowledge and enable them to distinguish new information: According to Ausubel's (2000) theory of meaningful learning, new information is learned by associating it with previous knowledge. In machine learning instruction, it is important to consider students' readiness levels when determining the starting point of instruction. For instance, machine learning deals with problems that cannot be solved by algorithms. To help students understand this concept, it is useful to present them with problem examples from the mathematics course that can be solved by algorithms. Additionally, the process of creating a model based on data can be explained by using the line equation between two points in analytical geometry, which students have previously learned.

- Use a question-based teaching approach: To motivate students to learn, present them with real-life, ill-structured problems. For instance, by asking why artificial intelligence or machine learning is necessary, you can create a discussion environment about the need for these technologies and methods.
- Foster a collaborative learning environment: In real life, teamwork is essential. To succeed in business, it is crucial for students to gain experience in cooperative projects, particularly those with low risk. When creating applications for artificial intelligence and machine learning, students should be encouraged to form groups and work together to solve problems.
- Allow students to determine their learning process: Teaching students how to learn is crucial for developing problem-solving skills. It is beneficial for students to determine their learning process while being supervised. This will support their learning. Once students have acquired the basic knowledge and skills related to machine learning, they should be provided with information about the data sources that can be used and allowed to create their modeling processes.
- Provide students with information about the resources they can use: Providing students with information about the cognitive tools and web resources available to them can facilitate their engagement in extracurricular activities. The knowledge and skills to be conveyed by teaching in a limited time will also be limited. Therefore, in addition to the guides on machine learning, students can be provided with information about reliable multimedia materials available on the web pages and students can be supported outside the course, especially in the project development process.
- Organise a presentation activity to encourage students' participation: Presenting the products they have prepared to their peers and individuals who are not knowledgeable about the subject will increase students' participation as well as improve their presentation skills. Transferring the applications prepared for machine learning to the student's peers and individuals who are interested in the subject by creating an exhibition area will provide feedback on the deficiencies in the students' applications.
- Create a timetable for the students: It will be useful to create a timetable for the design process so that students can carry it out with a sense of responsibility. In this way, students will be aware of the deadlines in the product development steps and will make the necessary preparations according to this schedule. In machine learning classes, steps such as identifying the problem, identifying data sources, identifying input and output variables, and building models can be required of students in specific weeks and a work schedule can be presented to students.

Table 5.1 Symbolic representation of the implementation process

Group	Pretest		Posttest
G	T1	X	T2

G: The group in which artificial intelligence teaching was carried out
T_1, T_2: Computational thinking skills scale
X: Design-based learning environment for teaching artificial intelligence

Method

This section presents examples of AI-based applications that integrate computational thinking and an application example for creating a design-based learning environment.

Research Model and Participants

The implementation sample was conducted as a single group pretest–posttest design. The symbolic representation of the research model in the implementation process is shown in Table 5.1.

Implication Process

Before the implementation process began, a 7-week teaching program was designed by the researcher. The topics and content of the curriculum are given below:

- Definitions and Basic Concepts (Week 1): This topic explained basic concepts related to artificial intelligence, machine learning, deep learning, and computer science. Although the students were trained in using computers before the application, it was observed during the application that the students had no experience in writing algorithms for problem-solving. Therefore, in addition to the concepts related to artificial intelligence within the topic of definitions and basic concepts, subtopics such as in which cases artificial intelligence and machine learning are necessary, examples of problems that can and cannot be written algorithms, the importance of artificial intelligence applications, the differences between the classical programming approach and machine learning were mentioned and the student's level of readiness about machine learning was ensured. To connect the students with their previous learning, the concept of algorithm was emphasized and examples of problems for which algorithms can be written in everyday life were explained. Then it was explained that students cannot directly write algorithms for many problems in real life and the differences between previous knowledge and new knowledge were explained. Different sorting algorithms were shown to the students and question–answer activities were carried out to explain why more than one algorithm can be written for a problem and how to check the effectiveness of these algorithms. At the end of the training, the students were presented with a work schedule for the machine learning activity they would perform at the end of the training, and it was stated that they should follow this work schedule every week and progress the process by forming groups of 2–3 people. Different e-books that students can use outside the course and resources on machine learning and Orange usage that they can access from the YouTube platform were introduced.

- How machines learn (week 2): How supervised, unsupervised, semi-supervised, and reinforcement learning approaches used in the machine learning process are explained to the students. At this point, the basic concepts used in machine learning such as data sets, data mining applications, input, and output variables, and labeled data within the learning approaches were introduced to the students. For the students to understand how machines learn, the input–output-rules process in machine learning is explained instead of the input-rules-output process in the classical approach by stating the differences between them. The course explained how to build a predictive model through a question–answer activity, using the example of not being able to write an algorithm to evaluate the house price in a region. One of the most important inputs in machine learning is data. As the creation of data takes time, the students were introduced to the Kaggle application and the Google Dataset search engine, which they can use in this step. In the final step of this week, students were divided into groups to ensure that they worked collaboratively, and students were asked to research examples of problems that can be solved with artificial intelligence in real life.
- The stages of machine learning and application examples (weeks 3–6): In this phase, the following topics were explained: Partitioning the data into learning and test data in the machine learning process, building the model according to the learning data, making predictions using the test data of the built models, comparing the predictions made for the test data with the actual data, and evaluating the model and the confusion matrix. The question–answer method was used to explain the topics. As machine learning algorithms, the working processes of kNN, Naive Bayes networks, and decision tree techniques were superficially explained to the students. In this step, the students perceived the differences between the algorithms and understood why more than one machine-learning algorithm was developed. During the weekly intervals in this work package, the students also had the opportunity to develop their applications in the laboratory environment. They did this in the company of their peers and the teacher.
- Presentation and discussion of case studies (week 7): The developed application examples were presented by the students to their peers for feedback and information sharing.

The next section provides information on the features and explanations of the Orange Data Mining application used in the study.

Orange Data Mining Application

Orange Data Mining is an open-source toolkit for data visualization, machine learning, and data mining (Demšar et al., 2013). It is a user-friendly tool that can be used to teach machine learning visually (Gresse von Wangenheim et al., 2021). Orange enables users to perform machine learning processes without requiring any coding experience, making it easy to learn and use. The following section describes the main components of the Orange application and their respective tasks:

- Data: This component allows for the inclusion of the file to be used in the machine learning application within the working environment. This enables the use of data sources for working purposes.
- Transform: This component enables data preprocessing steps, such as selecting specific columns or rows, transposing, grouping data, and assigning missing values, to be performed on the data added to the working environment. This prepares the data for analysis.
- Visualize: This component enables the visualization of data distributions and analysis results through visual elements such as decision trees and scatter plots.
- Model: This component prepares data for analysis using data mining methods. Various algorithms, including kNN, Tree, SVM, and Naive Bayes, can be used to create models. Additionally, models created with this component can be saved, or previously saved models can be added to the working environment.
- Evaluate: This component generates models that can be evaluated for accuracy rates and compared. The models can also be used to make predictions based on different inputs or to examine the distribution of correct predictions using the confusion matrix.
- Unsupervised: This component allows for the application of unsupervised machine-learning techniques, such as clustering algorithms. However, as these techniques were not covered in the teaching process, the students did not utilize them in their applications.

The Orange application includes different options within the mentioned components. Users can add required options to the working environment through drag-and-drop operations. Figure 5.2 illustrates an application created with students during teaching activities.

The items in Fig. 5.2 were added in order based on the machine-learning process step during the narration. For example, firstly, the File component was added to add a data source to the application and the data source file was selected with this component. Then, the variables and data set in the data source are shown with the Data Table component. The distributions component was added to examine the distribution of data according to categorical variables and the Scatter Plot component was added

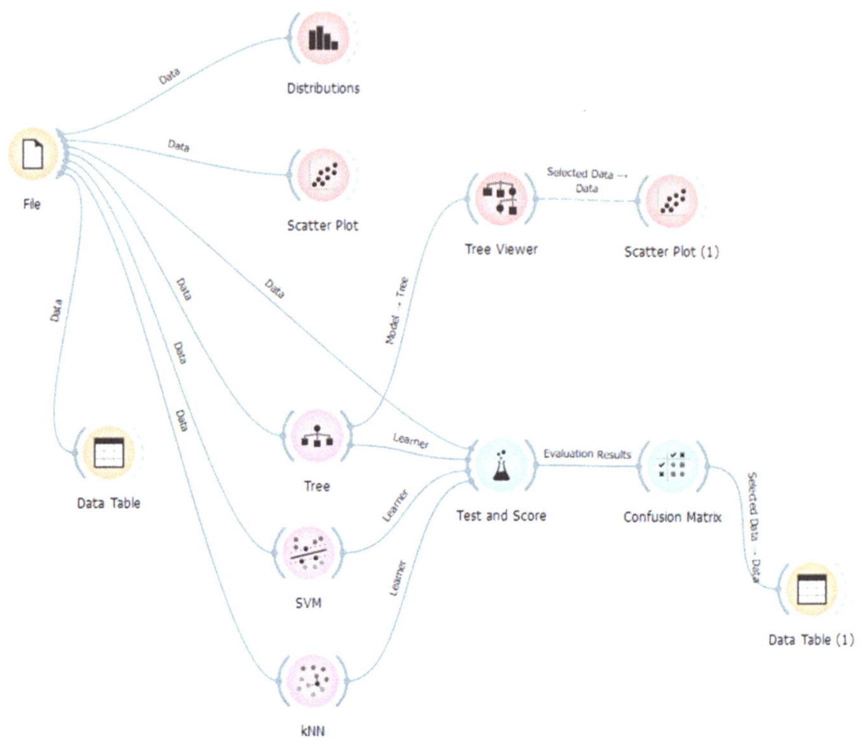

Fig. 5.2 Flow process example using Orange data mining

to the working environment to examine the distribution of continuous variables. In this example, the creation of models with Tree, SVM, and kNN algorithms and the comparison of the created models were carried out. With the Test and Score component, students can view the results of the comparison of the models. Confusion Matrix was added to display the number of incorrect or correct predictions in the models. A Data Table component has been added to display the rows in which the correct and incorrect predictions are in the Confusion Matrix. The Orange application enabled the students to interpret the outputs more easily since it allows the modelling to be done easily as well as displaying the graph or data set according to the selected criteria.

Data Collection Tools

The study utilized Korkmaz et al.'s (2017) computational thinking skills scale to assess changes in university students' computational thinking skills before and after the activities. The scale comprises 29 items rated on a 5-point Likert scale, with scores ranging from 1 (most negative) to 5 (most positive). The minimum score attainable on the scale is 29, while the maximum is 145.

The study employed a computational thinking skills scale with five dimensions: creativity, algorithmic thinking, collaboration, critical thinking, and problem-solving. In the scale development study by Korkmaz et al. (2017), the Cronbach alpha reliability coefficient for the entire scale was calculated as 0.822. The reliability coefficients for the creativity, algorithmic thinking, cooperativity, critical thinking, and problem-solving dimensions were calculated as 0.843, 0.869, 0.865, 0.784, and 0.727, respectively. The pretest and posttest total scores had Cronbach alpha reliability coefficients of 0.892 and 0.902, respectively. The pre-test scores yielded Cronbach alpha reliability coefficients of 0.724, 0.880, 0.926, 0.800, and 0.787 for creativity, algorithmic thinking, cooperativity, critical thinking, and problem-solving factors, respectively. The post-test scores yielded coefficients of 0.812, 0.866, 0.939, 0.820, and 0.674 for the same factors.

Findings and Discussion

Table 5.2 presents the results of t-test analyses on the mean scores of students' computational thinking skills before and after the activity.

There is no significant difference between the pre-test (M = 112.11) and post-test (M = 113.07) mean scores of the students on the computational thinking skills test. $t(34) = -0.581$, $p > 0.05$. From this result it can be said that the design-based teaching activities did not affect the development of computational thinking skills. Considering the range of scores (29–145) obtained from the computational thinking skills scale, it is a notable finding that the students obtained a score well above the average score (87) to be obtained from the scale.

The mean scores of the participants' pre-test and post-test results for the sub-dimensions of computational thinking skills are explained. Upon examining the mean scores of the students in the creativity dimension, a significant difference between the pre-test (M = 32.77) and post-test (M = 33.89) mean scores was observed, $t(34) = -2.062$, $p < 0.05$. There was no significant difference found between the pre-test and post-test mean scores of the students in algorithmic thinking skills ($t(34) = 0.078$, $p > 0.05$). Similarly, in the dimension of cooperativity, no significant difference was observed between the pre-test mean scores of the students before the activity (M

Table 5.2 Results of t-test for mean scores of pre-test and post-test computational thinking skills

Measurement (computational thinking skills)	n	M	S	df	t	*p*
Pre-test	35	112.11	13.45	34	– 0.581	0.57
Post-test	35	113.07	12.79			

Table 5.3 t-test results of the mean scores of the dimensions of computational thinking skills pre- and post-test

Factor	Min score	Max score	Measurement	n	M	S	df	t	*p*
Creativity	8	40	Pre-test	35	32.77	4.04	34	– 2.062	0.047
			Post-test	35	33.89	3.80			
Algorithmic thinking	6	30	Pre-test	35	22.09	4.24	34	0.078	0.939
			Post-test	35	22.04	3.65			
Cooperativity	4	20	Pre-test	35	15.46	3.60	34	– 0.413	0.682
			Post-test	35	15.63	3.91			
Critical thinking	5	25	Pre-test	35	18.14	3.60	34	0.062	0.951
			Post-test	35	18.11	3.37			
Problem solving	6	30	Pre-test	35	23.66	4.09	34	0.393	0.697
			Post-test	35	23.40	3.30			

= 15.46) and the post-test mean scores after the activity (M = 15.63) (t(34) = – 0.413, $p > 0.05$). There was no significant difference in the critical thinking dimension between the mean score of the students in the pretest (M = 18.14) and the mean score in the posttest (M = 18.11), t(34) = – 0.062, $p > 0.05$. Similarly, in the problem-solving dimension, there was no significant difference between the students' mean pretest score (M = 23.66) and mean posttest score (M = 23.40), t(34) = 0.393, $p > 0.05$.

The study's findings indicate that teaching artificial intelligence from the design-based learning environment perspective does not result in a statistically significant difference in students' computational thinking skills, except for creativity. The study's findings on the creativity dimension align with Saritepeci's (2020) research on the effectiveness of design-based learning activities in developing CT skills among high school students. However, there was no significant difference in the mean scores of students in the algorithmic thinking, cooperativity, critical thinking, and problem-solving dimensions.

The results regarding whether the scores obtained from the sub-dimensions of the scale, as well as computational thinking skills in general, differed according to the pre-test and post-test means are shown in Table 5.3.

AI-Based Activity Examples for CT Skills

This section gives some examples of applications that have been used with students during educational activities.

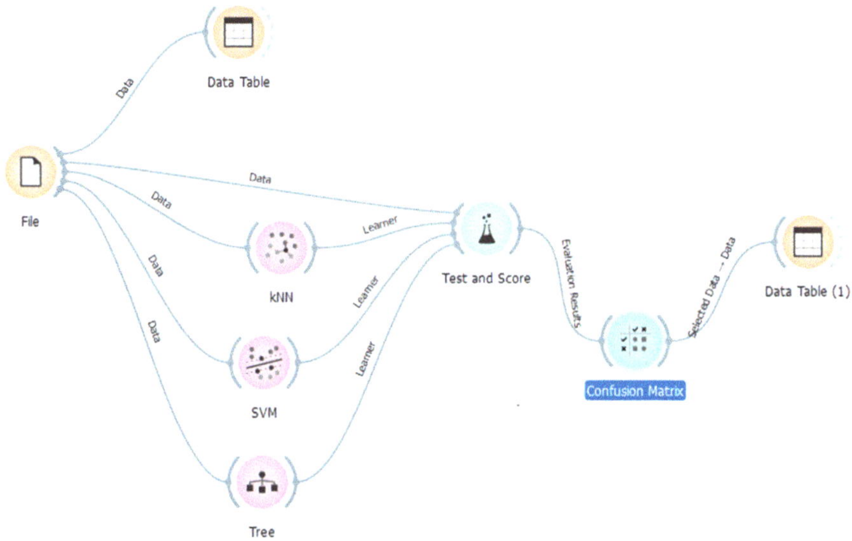

Fig. 5.3 Modelling process for estimating rainfall with Orange

Example 1: Predicting Rainfall Based on Weather-Related Values

In this example, the students wanted to build a model using some variables related to the weather and predict precipitation according to the model built. To do this, the students accessed the file Weather_Data.csv using the Kaggle web application and analyzed the variables. In this process, the students were first advised to create a logical framework when determining the input variables. Then the input and output variables were determined on the downloaded dataset and the modeling processes were performed using the Orange application. The visualization of the modeling process created by the students using Orange is shown in Fig. 5.3.

In this example, students tested the effectiveness of more than one algorithm and used the Confusion Matrix to interpret the models. They also included the Data Table component to re-examine the rows that were not correctly predicted by the Confusion Matrix.

Example 2: Estimation of Apple Quality

In this example, students aimed to create a model using the variables size, weight, sweetness, crunchiness, juiciness, ripeness, acidity, and quality of apples. Students accessed the file apple_quality.csv from the Kaggle application and identified the

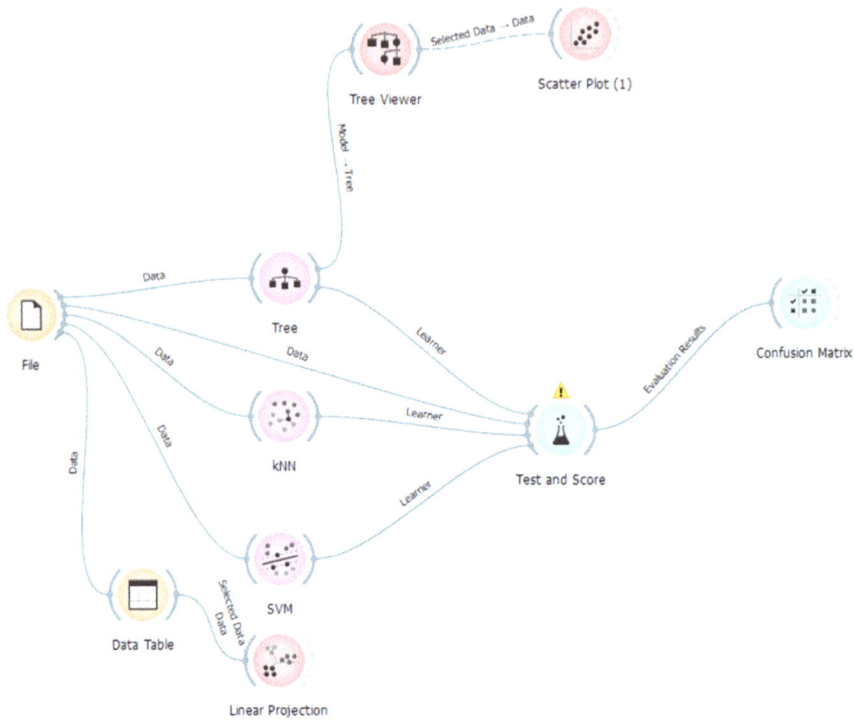

Fig. 5.4 Modelling process for predicting apple quality

input and output variables in the file. The visualization of the modeling is shown in Fig. 5.4.

The accuracy scores of the models produced by tree, kNN, and SVM were examined and the confusion matrices produced by these methods were analyzed. In this example, it is thought that the students created visualizations using the linear projection element, which was not taught in the course, and that the resources recommended outside the course were used.

Example 3: Classification and Prediction of Keçimen and Besni Grape Varieties

This activity aims to use the dataset (Raisin_Dataset.xlsx) obtained from Keçimen and Besni grape varieties, to apply different machine learning models to this dataset through the Orange Data Mining program and to compare the performance of these models. The visualization of the modeling process created in the Orange application is shown in Fig. 5.5.

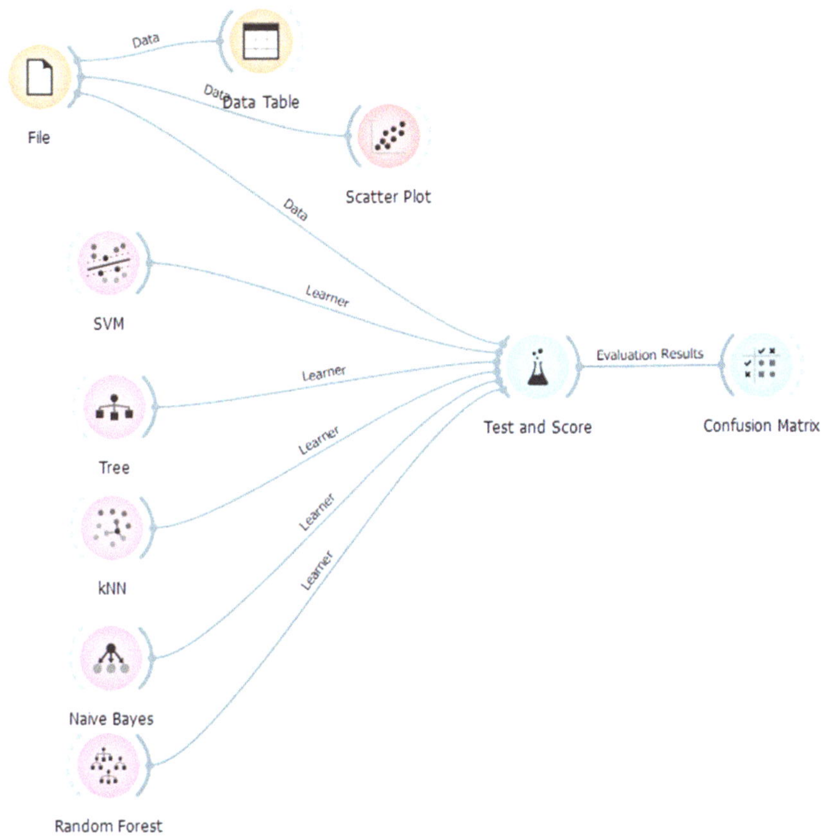

Fig. 5.5 Modeling process of Keçimen and Besni grape varieties

In this activity, the student investigated how successful he/she was in discriminating between Keçimen and Besni grape varieties using Support Vector Machines (SVM), Decision Trees, k-nearest Neighbour (kNN), Naive Bayes and Random Forest techniques.

Conclusion

This chapter provides an illustrative example of the implementation of artificial intelligence in the context of teaching activities designed according to the design-based learning approach to the development of computational thinking skills. In accordance with the findings of the present study, the following section will present the implementation for future practice and research and a summary of the key points.

Summary of Key Points

This section provides an illustrative example of the implementation of artificial intelligence in the context of teaching activities designed according to the design-based learning approach to the development of computational thinking skills.

Key points:

- Teaching pre-service teachers about how artificial intelligence tools work and how to use them in educational contexts is important for effective use of the tools.
- There is a relationship between computational thinking skills and the workings of artificial intelligence tools.
- Design-based learning environments can be used to teach the working processes of AI tools.
- It is more appropriate to use visual-based data mining applications, especially for people with low knowledge of the basics of artificial intelligence.

Implementations for Future Practice and Research

The main limitation of this study is that the pre-test scores of the participants' computational thinking skills were high and no other participants could be included in the study. In future studies, it is planned to investigate the effectiveness of the activities by selecting a sample group whose pre-test score is close to the mean score to be obtained from the scale.

In this study, taking into account the students' computer skills, they were asked to create models from data sets using the Orange application and to compare and interpret the models created. As the students could not automate the process specified in the last step of machine learning with the Orange application, the impact of machine learning application development processes using text-based programming languages on the development of computational thinking skills can be investigated in future research.

Although there was no significant difference in the change in students' computational thinking skills at the beginning and end of the study, it was observed that students' active participation in the learning process was ensured through design-based activities throughout the study process. The following are strategies that practitioners should consider when preparing design-based activities, considering the principles of project-based learning (Condliffe, 2017) and problem-based learning processes (Nilson, 2016):

- Relate to previous knowledge and enable them to recognize new information.
- Promote the learning process through questions.
- Provide a collaborative learning environment.
- Allow students to control their learning process.
- Provide students with information about the resources available to them.
- Organize a presentation activity to encourage student participation.

- Create a work plan for the students.

Considering the initial scores of the participants in the study, it can be seen that their computational thinking skills are above average in the range of scores obtained from the scale. Therefore, it is thought that it would be useful to investigate the effect of design-based activities for teaching artificial intelligence to students with low levels of computational thinking skills.

References

Alpaydin, E. (2021). *Machine learning*. MIT Press.

Ausubel, D. P. (2000). *The acquisition and retention of knowledge: A cognitive view*. Springer Science & Business Media.

Aydoğdu, Ş. (2020). Blok Tabanlı Programlama Etkinliklerinin Öğretmen Adaylarının Programlamaya İlişkin Öz Yeterlilik Algılarına ve Hesaplamalı Düşünme Becerilerine Etkisi [The effect of block-based programming activities on pre-service teachers' computer programming self-efficacy and computational thinking skills]. *Eğitim Teknolojisi Kuram ve Uygulama, 10*(1), 303–320. https://doi.org/10.17943/etku.649585

Azizan, S. A., & Abu Shamsi, N. (2022). Design-based learning as a pedagogical approach in an online learning environment for science undergraduate students. *Frontiers in Education.*

Chen, L., Chen, P., & Lin, Z. (2020). Artificial intelligence in education: A review. *IEEE Access, 8*, 75264–75278.

Condliffe, B. (2017). Project-based learning: A literature review. Working Paper. *MDRC*.

Demšar, J., Curk, T., Erjavec, A., Gorup, Č., Hočevar, T., Milutinovič, M., Možina, M., Polajnar, M., Toplak, M., & Starič, A. (2013). Orange: Data mining toolbox in Python. *The Journal of Machine Learning Research, 14*(1), 2349–2353.

Dong, W., Li, Y., Sun, L., & Liu, Y. (2024). Developing pre-service teachers' computational thinking: A systematic literature review. *International Journal of Technology and Design Education, 34*(1), 191–227. https://doi.org/10.1007/s10798-023-09811-3

Ferrara, S., & Qunbar, S. (2022). Validity arguments for AI-based automated scores: Essay scoring as an illustration. *Journal of Educational Measurement, 59*(3), 288–313.

Gabriele, L., Bertacchini, F., Tavernise, A., Vaca-Cárdenas, L., Pantano, P., & Bilotta, E. (2019). Lesson planning by computational thinking skills in Italian pre-service teachers. *Informatics in Education, 18*(1), 69–104.

Géron, A. (2022). *Hands-on machine learning with Scikit-Learn, Keras, and TensorFlow*. O'Reilly Media, Inc.

Gligorea, I., Cioca, M., Oancea, R., Gorski, A.-T., Gorski, H., & Tudorache, P. (2023). Adaptive learning using artificial intelligence in e-learning: A literature review. *Education Sciences, 13*(12), 1216.

Gresse von Wangenheim, C., Hauck, J. C. R., Pacheco, F. S., & Bertonceli Bueno, M. F. (2021). Visual tools for teaching machine learning in K-12: A ten-year systematic mapping. *Education and Information Technologies, 26*(5), 5733–5778. https://doi.org/10.1007/s10639-021-10570-8

Hsu, T.-C., Chang, S.-C., & Hung, Y.-T. (2018). How to learn and how to teach computational thinking: Suggestions based on a review of the literature. *Computers & Education, 126*, 296–310. https://doi.org/10.1016/j.compedu.2018.07.004

Kabudi, T., Pappas, I., & Olsen, D. H. (2021). AI-enabled adaptive learning systems: A systematic mapping of the literature. *Computers and Education: Artificial Intelligence, 2*, 100017.

Kaspersen, M. H., Bilstrup, K.-E. K., & Petersen, M. G. (2021). The machine learning machine: A tangible user interface for teaching machine learning. In *Proceedings of the fifteenth international conference on tangible, embedded, and embodied interaction, Salzburg, Austria.* https://doi.org/10.1145/3430524.3440638

Korkmaz, Ö., Çakir, R., & Özden, M. Y. (2017). A validity and reliability study of the computational thinking scales (CTS). *Computers in Human Behavior, 72*, 558–569. https://doi.org/10.1016/j.chb.2017.01.005

Nilson, L. B. (2016). *Teaching at its best: A research-based resource for college instructors.* John Wiley & Sons.

Ojo, O. (2024). *Promoting artificial intelligence education in K–12 through pre-service teachers' development* Itä-Suomen yliopisto.

Rajapakse Mohottige, N. U. S., Bjerke, A. H., & Andersen, R. (2024). Teacher education as stakeholder: Teacher educator perspectives on the integration of computational thinking into mathematics and science courses. *Teachers and Teaching*, 1–15. https://doi.org/10.1080/13540602.2024.2313635

Ramesh, D., & Sanampudi, S. K. (2022). An automated essay scoring systems: a systematic literature review. *Artificial Intelligence Review, 55*(3), 2495–2527. https://www.ncbi.nlm.nih.gov/pmc/articles/PMC8460059/pdf/10462_2021_Article_10068.pdf

Sands, P., Yadav, A., & Good, J. (2018). Computational thinking in K-12: In-service teacher perceptions of computational thinking. In M. S. Khine (Ed.), *Computational thinking in the STEM disciplines: Foundations and research highlights* (pp. 151–164). Springer International Publishing. https://doi.org/10.1007/978-3-319-93566-9_8

Sanusi, I. T. (2021). Teaching machine learning in K-12 education. In *Proceedings of the 17th ACM conference on international computing education research, virtual event, USA.* https://doi.org/10.1145/3446871.3469769

Sanusi, I. T., Olaleye, S. A., Oyelere, S. S., & Dixon, R. A. (2022). Investigating learners' competencies for artificial intelligence education in an African K-12 setting. *Computers and Education Open, 3*, 100083. https://doi.org/10.1016/j.caeo.2022.100083

Sanusi, I. T., Oyelere, S. S., Vartiainen, H., Suhonen, J., & Tukiainen, M. (2023). A systematic review of teaching and learning machine learning in K-12 education. *Education and Information Technologies, 28*(5), 5967–5997. https://doi.org/10.1007/s10639-022-11416-7

Saritepeci, M. (2020). Developing computational thinking skills of high school students: Design-based learning activities and programming tasks. *The Asia-Pacific Education Researcher, 29*(1), 35–54. https://doi.org/10.1007/s40299-019-00480-2

Sarkar, R., & Ahammed, S. P. (2023). Enhancing competency of teachers through artificial intelligence. *Journal of Arts, Humanities and Social Sciences.*

Selby, C., & Woollard, J. (2014). *Refining an understanding of computational thinking.*

Shamir, G., & Levin, I. (2022). Teaching machine learning in elementary school. *International Journal of Child-Computer Interaction, 31*, 100415. https://doi.org/10.1016/j.ijcci.2021.100415

Sundberg, L., & Holmström, J. (2024). Teaching tip: Using no-code AI to teach machine learning in higher education. *Journal of Information Systems Education, 35*(1). http://urn.kb.se/resolve?urn=urn:nbn:se:umu:diva-207861

Tedre, M., Toivonen, T., Kahila, J., Vartiainen, H., Valtonen, T., Jormanainen, I., & Pears, A. (2021). *Teaching machine learning in K-12 computing education: Potential and pitfalls.* arXiv preprint arXiv:2106.11034.

Umutlu, D. (2022). An exploratory study of pre-service teachers' computational thinking and programming skills. *Journal of Research on Technology in Education, 54*(5), 754–768.

Vartiainen, H., Tedre, M., & Valtonen, T. (2020). Learning machine learning with very young children: Who is teaching whom? *International Journal of Child-Computer Interaction, 25*, 100182. https://doi.org/10.1016/j.ijcci.2020.100182

Villalustre, L., & Cueli, M. (2023). Assessing the computational thinking of pre-service teachers: A gender and robotics programming experience analysis. *Education Sciences, 13*(10), 1032.

Voon, X. P., Wong, S. L., Wong, L. H., Khambari, M. N. M., & Syed-Abdullah, S. I. S. (2023). Developing pre-service teachers' computational thinking through experiential learning: Hybridisation of plugged and unplugged approaches. *Research and Practice in Technology Enhanced Learning, 18*, 006–006.

Wang, Y. (2021). An improved machine learning and artificial intelligence algorithm for classroom management of English distance education. *Journal of Intelligent & Fuzzy Systems, 40*(2), 3477–3488.

Wing, J. M. (2006). *Computational thinking* (Vol. 49). Association for Computing Machinery. https://doi.org/10.1145/1118178.1118215

Yadav, A., Stephenson, C., & Hong, H. (2017). Computational thinking for teacher education. *Communications of the ACM, 60*(4), 55–62. https://doi.org/10.1145/2994591

Zhang, F., Markopoulos, P., & Bekker, T. (2018). The role of children's emotions during design-based learning activity: a case study at a Dutch high school. In *10th International conference on computer supported education (CSEDU 2018).*

Zhang, F., Markopoulos, P., & Bekker, T. (2020). Children's emotions in design-based learning: A systematic review. *Journal of Science Education and Technology, 29*(4), 459–481. https://doi.org/10.1007/s10956-020-09830-y

Zhang, Q., Lu, J., & Jin, Y. (2021). Artificial intelligence in recommender systems. *Complex & Intelligent Systems, 7*, 439–457.

Chapter 6
Integrating Computational Thinking into Computer-Assisted Language Learning: From Theoretical Foundations to Design-Based Learning Applications

Ali Duran

Abstract This chapter provides an overview of the transformative potential of integrating computational thinking (CT) with computer-assisted language learning (CALL) within the field of foreign language education. This chapter, in particular, begins with a historical overview of the progression of technology, particularly the advent of computers, in the domain of language education, tracing how the integration of information and communication technology (ICT) has evolved to enhance language learning experiences. This chapter offers insights into the concept of CALL, defined as the process of learning languages in any setting through the use of computer technologies, and discusses its significance in creating enriched language learning environments. This chapter draws on a critical examination of three prominent perspectives that have surfaced in relation to the synergy between language learning theories, general learning theories, and their interplay with ICT: the structural, the cognitive/constructivist, and the sociocognitive. Through this examination, it is argued that the integration of design-based learning (DBL) and CT within language education is a contemporary manifestation of socio-cognitive theory, providing a dynamic and interactive learning experience that equips learners for real-world communication. The chapter further introduces CT-driven DBL activities as a novel approach to incorporating CT into language education, challenging misconceptions about the complexity of teaching CT. A detailed discussion on the steps of the standard design process and the concept of backwards thinking in DBL is presented, underscoring the importance of the symbiosis of cognitive engagement and physical action in the cultivation of ideas and solutions and establishing a foundation for effective and immersive language learning practices. Finally, this chapter concludes with theoretical insight and practical guidance for educators seeking to harness the power of CALL and CT in language education, paving the way for innovative practices through DBL.

A. Duran (✉)
Department of Foreign Language Education, Amasya University, Amasya, Türkiye
e-mail: aliduran@amasya.edu.tr

M. Saritepeci and H. Yildiz Durak (eds.), *Integrating Computational Thinking Through Design-Based Learning*, https://doi.org/10.1007/978-981-96-0853-9_6

Keywords Computer-assisted language learning · Computational thinking · Design-based learning · Information and communication technology · Socio-cognitive theory

Introduction

Technology is rapidly changing the world, necessitating adjustments in education to teach individuals the skills required in the fast-paced digital life (Rottenhofer et al., 2021; Wing, 2006). One problem-solving skillset that has gained considerable attention in the last couple of years is CT, first presented by the educationalist and mathematician Papert (1990), who argued for mathematics, and indeed all science, to be taught in new ways based on computation (Curzon & McOwan, 2017). This skillset has since gained increasing attention, found its place in many national curricula, and its definition, teaching, and evaluation have been discussed by various scholars (Grover & Pea, 2013; Wing, 2006). Currently, education stakeholders and organizations worldwide have recognized the importance of CT for young generations, with a rapidly growing passion for cultivating students' CT skills at different levels of society, including academia, policy, and mass participation (Grover & Pea, 2013).

CT was classified into 11 thinking processes by Wing (2006). However, Hsu et al. (2018) classified 19 different thinking processes as a result of a review of studies on CT between 2006 and 2017. One of the important areas where CT has an impact is undoubtedly language learning and teaching (Jacob et al., 2018, 2022; Mubarok et al., 2023). The convergence of technology and education has opened new frontiers in teaching methodologies, particularly in language learning (Lu & Fletcher, 2009). This chapter delves into the integration of CT within the study of language teaching, emphasizing a design-based approach. As noted by Rottenhofer et al. (2021) and Sabitzer (2012, 2013), CT has emerged as a crucial skill set, transcending its traditional confines within computer science to influence various educational disciplines, including language learning. CT's emphasis on problem-solving, pattern recognition, algorithmic thinking, and abstraction equips learners with a versatile toolkit for tackling complex language-related challenges (Wolz et al., 2011). In this sense, the objective of this chapter is to explore the theoretical and practical aspects of incorporating CT into design-based language learning (DBLL).

This chapter examines the intersection of computational thinking (CT) and computer-assisted language learning (CALL) within foreign language education, illuminating the transformative potential of their integration. The chapter starts with a historical overview of how technology, especially computers, has played a pivotal role in language education. This journey traces the evolution of information and communication technology (ICT) and its growing significance in enhancing language learning experiences. CALL is defined as the process of learning languages in any setting through the use of computer technologies, emphasizing its importance in creating rich and interactive language learning environments. This exploration is grounded in a critical examination of three perspectives that have emerged at the confluence of language learning theories, general learning theories, and their interaction with ICT: the structural, the cognitive/constructivist, and the sociocognitive. Through this analysis, it is argued that integrating DBL and CT within language education represents a modern reflection of socio-cognitive theory. This integration provides a dynamic and interactive learning experience, equipping learners with the skills necessary for real-world communication. CT-driven DBL activities are introduced as an innovative approach to embedding CT into language education. This approach challenges the misconceptions surrounding the complexity of teaching CT. The steps of the standard design process and the concept of backward thinking in DBL are discussed in detail, emphasizing the symbiosis of cognitive engagement and physical action. This combination is crucial for fostering the development of ideas and solutions, thus establishing a solid foundation for effective and immersive language learning practices. This chapter is based on the premise that readers should have a comprehensive understanding of how CT can transform language learning, making it more engaging, effective, and aligned with the demands of the modern world by the end of this chapter. Finally, the chapter concludes with theoretical insights and practical guidance for educators looking to exploit the synergies between CALL and CT in language education. The goal is to pave the way for innovative practices through DBL, advocating for a shift towards more interactive, engaging, and efficacious language learning methodologies.

Integrating CT into Language Learning Through DBL

Computer-Assisted Language Learning (CALL): A Historical and Theoretical Perspective

Early Developments in CALL

To better grasp the significance of DBL in incorporating CT into language education (Garrett, 2009), it would be better to first consider the historical progression of technology, particularly computers, within this field. This historical context will illuminate how the integration of CT and DBL into contemporary language education theories has evolved. By examining the developmental stages of technology's use in language teaching, it will be possible to have a better picture of the role and impact of these modern educational approaches. This exploration helps to frame the ongoing dialogue about educational innovation and the ways in which technology can further enrich language learning experiences.

Since the early 1960s, the approach to language teaching has undergone significant transformations. The emphasis has shifted from merely teaching grammatical rules to enhancing communicative skills (Kern & Warschauer, 2000). The evolution in teaching methodologies has led to an increased emphasis on understanding language in a comprehensive manner, highlighting the importance of using language as a tool for communication rather than just memorizing its structural aspects (Ellis, 1999). This shift has paved the way for innovative teaching techniques that prioritize meaningful interaction and the practical use of language in real-life contexts (Armstrong & Ferguson, 2010). The importance of negotiating meaning outweighs the repetitive nature of structural drills (Pica, 1996). The focus on understanding has grown, making the delivery of understandable language content a key teaching goal (Echevarria et al, 2008). Culture has gained attention and importance, despite some educators' uncertainties about how to effectively incorporate it into language teaching (Nieto, 2001). Modern language textbooks now differentiate between spoken and written language, incorporating authentic materials and literary works (Gilmore, 2007). This period of diverse changes has seen the maturation of CALL as a pivotal innovation in language education (Kern & Warschauer, 2000). On the other hand, the significance of names extends beyond mere description of concepts, influencing our understanding of these concepts. Initial terminologies used in the context of language acquisition through technology, like "computer-aided instruction (CAI)", highlight a focus on the technological aspect and a behaviorist perspective towards learners (Healey, 2016). The introduction of microcomputers in the 1980s facilitated increased computer integration into educational settings. Despite the prevalence of CAI-style drill and practice, occasionally termed "computer-assisted language instruction (CALI)", some of the initial educators who incorporated computers into teaching advocated for a shift in terminology to specifically focus on language learning (Clifford, 1987). The acronym 'CALL' for computer-assisted or computer-aided language learning has become the widely accepted term (Davies & Higgins, 1982).

Integration of ICT

Within this transformative landscape, the integration of information and communication technology (ICT) in language learning has bridged the gap between theoretical frameworks and practical application (Alkaromah et al., 2020). This convergence has facilitated a more dynamic and interactive learning environment, fostering deeper engagement and understanding among learners (Gulruh et al., 2020). Three prominent perspectives have emerged and been explored in relation to the connection between language learning theories, general learning theories, and their interaction with ICT (Kern & Warschauer, 2000): the structural, the cognitive/constructivist, and the sociocognitive.

The structural theory in language education emphasized mastering language systems and structures to produce grammatically correct language. This approach spanned several decades and included methods such as grammar-translation, the audio-lingual method, and the use of repetitive drills for habit formation, as well as contrastive analysis (Kern & Warschauer, 2000). This era's focus on precision and repetition in language learning laid the groundwork for the initial development of CALL tools. These tools were primarily aimed at reinforcing linguistic accuracy through repetitive practice, mirroring the drill-based methodologies of the time. However, the reliance on presenting only right or wrong answers limited the depth of learning and understanding (Beatty, 2013). In the field of CALL, this emphasis manifested in software designed to enhance formal linguistic accuracy via systematic practice. Such programs, often resembling digital textbooks, offered limited feedback options, distinguishing solely between correct and incorrect responses. This approach, largely aligning with the behaviorist pedagogy and technology advancements from the 1950s to the 1970s, eventually faced criticism from its intended end-users—language learners (Warschauer, 1996).

The cognitive/constructivist theory in language education emerged in response to the behaviorist approach to language acquisition. Chomsky (1965) introduced the concept of "transformational-generative grammar," emphasizing that language development is driven by inherent cognitive frameworks rather than the repetition and imitation of accurate linguistic patterns. Consequently, language acquisition began to be seen as a dynamic endeavor involving the creation and modification of knowledge (Warschauer et al., 2000). Developed in the late 1970s, Krashen's Monitor Model proposes a clear differentiation between acquisition and learning, suggesting that acquisition occurs subconsciously, whereas learning is a conscious activity, characterized by a state of awareness about the language (Krashen, 1981). This paradigm shift led to a decreased emphasis on accuracy, favoring approaches that incorporate elements like problem-solving and collaborative tasks, viewing language learning as a series of process-oriented activities (Beatty, 2013). This evolution is mirrored in CALL, where greater importance is given to the learner's engagement in activities, perceiving learning as an inventive process initiated through problem-solving or hypothesizing (Kern & Warschauer, 2000). During this period, critics argued that interactivity in computer systems was inadequate, and such systems lacked the intelligence to manage the diversity of interactions expected in communication (Cameron, 1989). This phase, spanning the late 1970s to the early 1980s, was also known as communicative CALL. This method emphasized discovery, expression, and development as key learning processes (Davies & Fitzpatrick, 2003) and was closely tied to cognitive theories. The constructivist/integrative CALL approach considered learners as active creators, originating from social or socio-cognitive learning theories (Lee, 2007). It aimed to merge language skills and technological tools in education, highlighting project-based, task- and content-driven learning in real-life contexts (Davies & Fitzpatrick, 2003). The socio-cognitive view shifted the emphasis towards viewing language learning as a socially constructed activity (Warschauer et al., 2000). Concepts such as communicative competence, social interaction, discourse, and learner strategies were developed and articulated during this time (Kramsch & Thorne, 2003).

The socio-cognitive theory in language education texts as communicative acts, as acting with words in meaningful interaction (Cheung et al., 2021; Kern & Warschauer, 2000). It suggests that communicative practices carry inherent and explicit ideas and prerequisites for interaction (Stapa, 2007). With the increasing integration of technology, the landscape for interaction within multimodal settings is evolving, altering the dynamics of communication (Khatib & Shakouri, 2013). This evolution necessitates a reevaluation of the limitations and advantages, along with strategies for interaction and collaboration, to understand when and how to effectively employ these modes for specific purposes (Shetzer & Warschauer, 2000). Consequently, this reasoning prompts a deeper examination of how learning activities are designed and learning goals are set, which becomes crucial for comprehending communication and interaction within these frameworks (Vigmo, 2010).

Initially, CALL was conceptualized as exploring and examining the role of computer applications in the instruction and acquisition of languages (Levy, 1997, p. 1). However, with the proliferation of various ICTs and the widespread integration of digital tools and resources in language education both within and beyond the classroom, CALL's definition evolved to encompass the creation and application of technological tools in language education (Levy & Hubbard, 2005, p. 143). In today's world, CALL encompasses a wide array of research focuses, including but not limited to corpus studies, cultural aspects, discourse analysis, remote education, student perceptions, genre analysis, digital literacies, user interface design, auditory comprehension, language literacy, preservation of less common languages, multimedia use, phonetics, textual comprehension, second language learning, societal influences, oral communication, speech recognition technology, linguistic structure, evaluative measures, lexicon development, online pedagogy, and composition. It is clear from global research that this field is characterized by its heterogeneity, linking traditional linguistic competencies (such as reading, listening, speaking, and writing) with a broader exploration of novel dynamics introduced by digital technology (Vigmo, 2010, p. 35). A more comprehensive perspective defines CALL as the process of learning languages in any setting through the use of computer technologies (Egbert, 2005), highlighting the use of any form of computer technology within a language learning environment (Barr & Gillespie, 2003). Unlike the domain of computer-supported or computer-assisted language learning (CALL), which has undergone extensive research over the past few decades with the advent of e-learning and technology-enhanced learning, the incorporation of CT and computer science principles into language teaching remains relatively rare (Sabitzer et al, 2018). However, this emerging intersection between CT, computer science, and language learning promises to revolutionize the way languages are taught and learned by fostering analytical thinking, problem-solving skills, and a deeper understanding of language structure and use through algorithmic thinking and programming concepts (Hsu et al., 2022; Rottenhofer et al., 2022).

The rise of internationalization and advancements in technology, along with transformations in societies and professional spheres, have altered the contexts in which languages operate and where learning occurs (Bergroth-Koskinen & Seppala, 2012). Placing learners within a social learning context where they engage with peers to acquire new information fosters the creation of a learning community. This setting enables them to observe, interact, and work collaboratively with others in order to jointly construct new understanding (Fillmore, 1991). In this sense, understanding the historical context of technology's role in language education and the evolution of teaching methodologies offers a comprehensive backdrop against which the importance of DBL in incorporating CT can be assessed. The transition from a structural to a socio-cognitive approach in language teaching, emphasizing communication, interaction, and the social construction of knowledge, lays the groundwork for integrating

DBL and CT into language education (Saad & Zainudin, 2022). This integration is not merely about using technology for the sake of it but rather about leveraging these approaches to enhance the learning experience in a manner that aligns with contemporary educational theories (Hung, 2017). DBL, with its focus on solving real-world problems through design and iteration, naturally complements the socio-cognitive theory's emphasis on social interaction and the construction of knowledge (Li, 2008). By incorporating CT into DBL, learners can develop a systematic way of thinking about problems and solutions, which is crucial for language learning (Egbert, 2005). CT's emphasis on decomposition, pattern recognition, abstraction, and algorithmic thinking can help learners break down language concepts into smaller, manageable parts, identify patterns in language use, abstract from specific instances to general principles, and develop step-by-step strategies for communication (Zhao et al., 2024). The relationship between socio-cognitive theory in language education and DBL in incorporating CT is, therefore, synergistic (Vogel et al., 2017). Socio-cognitive theory underscores the importance of context, interaction, and the social construction of knowledge in learning (Gunawardena et al., 2023). DBL, enhanced by CT, provides a structured yet flexible framework that allows learners to engage with language learning in a way that is interactive, contextually meaningful, and aligned with how knowledge is constructed socially (Chiu et al., 2023). Through DBL projects that incorporate CT, learners can work collaboratively on language-related tasks, using technology as a tool for exploration, expression, and connection (Perrenet et al., 2000). This approach not only makes learning more relevant and engaging but also mirrors the socio-cognitive perspective by situating language learning within the context of social interaction and collaborative problem-solving (Stewart et al., 2019; Swain & Watanabe, 2013). In essence, the integration of DBL and CT in language education represents a contemporary application of socio-cognitive theory, offering a dynamic and interactive learning experience that prepares learners for the complexities of real-world communication (Moreno & Kilpatrick, 2018; Reeves & McKenney, 2013).

CT-Driven DBL Activities for Language Learning: Theoretical Foundations and Practical Applications

Recent advancements in the field indicate a transition focusing more on the students and their language learning activities, diverging from the previously intense focus on the application of technology in terms of CALL (Kern, 2006). The definition and essence of CALL have undergone significant transformation (Thomas, 2013), reflecting an evolution in pedagogical approaches and technological integration within language education (Beatty, 2013; Van den Branden et al., 2009). The increasing emphasis on viewing the language learner as engaged and proactive within social and cultural environments is mirrored by concurrent advancements in interdisciplinary research that emphasizes cooperative learning through computer technology (Stahl et al., 2014). Embedding language within a socio-cultural framework, Halliday (1971) posits that language should be perceived as possessing a potential for meaning, conceptualized in terms that highlight what an individual is capable of achieving through linguistic expression.

Constructivist and Constructionist Approaches

Adopting a constructivist approach for teaching and developing constructivist learning settings is seen as a promising strategy for effectively blending CT and DBLL into language education (Csizmadia et al., 2019; Kirwan et al., 2018). Constructivist learning environments, facilitated by technology, encourage students to engage in critical thinking, problem-solving, and collaborative learning, which are essential skills for CT and DBLL (Rosa, 2016; Vigmo, 2010). On the other hand, there is another term used for CT integration into DBL, called constructionism (Parmaxi & Zaphiris, 2016). Constructionism, which builds upon the ideas presented by Piaget (1954), suggests that learners construct their own knowledge, rather than receiving it directly from an expert, such as a teacher. Constructionism, similar to constructivism, views learners as active participants in the creation of their own understanding. However, the former emphasizes the importance of learners creating tangible, shared artifacts (Ackermann, 2001). Unlike Piaget (1954), who concentrated on the cognitive processes involved in learning, constructionism, proposed by Papert (1980, 1993), highlights the value of hands-on learning through the creation of artifacts, often facilitated by digital and computer-based technologies. Constructionism emphasizes the process of learning through creation, highlighting the importance of the interactions between individual learners and the artifacts they create, often with the support of digital media and computer-based technologies (Kafai & Resnick, 1996). It can be argued that this is exactly what CT-driven DBLL activities are.

Such environments support language learners in applying their linguistic knowledge in meaningful, real-world tasks, thereby enhancing their language proficiency and digital literacy concurrently (Rosa, 2016). This dual focus not only advances learners' language skills but also prepares them for the demands of the digital age, making the constructivist approach a favorable method for integrating technology into language education (Kwan & Wong, 2015). Estes (2004) contends that traditional approaches to language education predominantly focus on didactic instruction and teacher-directed activities. Learning, however, is not a simple process of knowledge transmission (Jonassen & Land, 2012). Conley (2014) further emphasizes that learning within a forward-thinking educational framework is an active process rather than a passive one. Cohen (1999, p. 3) describes learners as active problem solvers and thinkers, constructing understanding through their unique experiences within both physical and cultural contexts. Wing (2006) posits that CT is a crucial skill, on par with literacy and numeracy, and advocates for its incorporation into everyday school activities and curricula (Sabitzer et al., 2018).

Previous research has reported beneficial effects from incorporating CT into language learning. For instance, certain studies have successfully integrated CT into writing workshops, resulting in significant improvements in students' ability to recognize emotions towards CT (e.g., Wolz et al., 2011). Consequently, it becomes imperative to view CT as an intrinsic component of students' analytical capabilities and to ensure its instruction to all children, irrespective of their focus on literacy, numeracy, or writing activities (Wing, 2006). Analogous to the way fundamental language skills facilitate communication, elementary CT skills enhance individuals' capacity to analyze information and complete tasks (Lu & Fletcher, 2009). To date, many countries have integrated CT into their educational syllabi, though misconceptions persist, particularly the belief that CT instruction demands extensive technical skills and deep computer science knowledge. Contrary to this belief, CT's applicability extends beyond technological subjects, empowering children to approach problems with strategies derived from computer science, without necessarily focusing on technology (Rottenhofer et al., 2021). In this sense, promoting CT extends beyond technology. CT emphasizes developing skills that enhance systematic and efficient problem-solving and task management (Barr et al., 2011).

Benefits of CT-Driven DBL Activities

CT-Driven DBL activities for language learning emerge as a promising solution to the integration of CT in language education, by addressing misconceptions regarding the complexity of teaching CT (Costa et al., 2016; Pande & Bharathi, 2020; Roy & Brine, 2013). These activities can significantly enhance language learning by fostering a hands-on, problem-solving approach that mirrors the process of learning a new language (Amazian et al., 2022; Cleminson & Cowie, 2021). Through CT-Driven DBL activities, students can engage in tasks that require them to analyze language patterns, construct and deconstruct sentences, and apply logical reasoning to understand grammatical structures, much like how they would tackle problems in computer science (Crites & Rye, 2020). This method not only aids in the development of CT skills but also deepens their understanding of the language by encouraging them to view language learning through the lens of problem-solving and design.

Moreover, CT-Driven DBL activities in language learning can democratize access to CT education, by proving that it does not necessitate extensive technical knowledge or resources. Instead, it leverages existing language learning frameworks to incorporate CT principles, making it more accessible and less intimidating for both students and educators (Hsu & Liang, 2021; Rottenhofer, 2022; Wu et al., 2024; Youjun & Xiaomei, 2022). Furthermore, such activities promote engagement and motivation among students, as they find relevance and application of CT in everyday learning tasks, making the learning process more interactive and enjoyable (Alonso-Prieto, 2019). By incorporating CT-Driven DBL activities into language learning, educators can foster critical thinking, problem-solving skills, and digital literacy, preparing students for a future where these skills are paramount (Rodríguez, 2022).

Design-Based Learning Process

Steps of the Standard Design Process

DBL represents a teaching methodology rooted in problem-based learning, where students aim to innovate and create concrete, tangible outputs—ranging from artifacts to systems and solutions—through the application of acquired knowledge (Gómez-Puente et al., 2013). DBL involves presenting students with real-world challenges to solve, thereby allowing them to engage actively in their learning journey by applying and reflecting on their acquired knowledge (Mehalik et al., 2008). Given that problem-solving is a fundamental aspect of daily life, it is argued that it should underpin both educational frameworks and societal structures (Ale, 2020). In this framework, learners are envisioned not only as idea generators and implementers but also as active contributors to their educational environments (Ale, 2020). While DBL has primarily been applied in fields like science, design, and engineering (Apedoe et al., 2008), it also offers potential for expansion into the social sciences, wherever there is a problem that can benefit from applied knowledge to generate tangible outcomes (Ale, 2020). DBL is supported by evidence advocating for the use of formative feedback within team-based projects as a means of assessment, focusing on both the process and the final product (Mehalik et al., 2008).

The primary focus of design lies in the creation and actualization of novel entities. This includes valuing "material culture" as well as engaging in the "art of strategizing, crafting, constructing, and executing". Fundamentally, it is rooted in the vernacular of shaping and forming. There is an opportunity to nurture students' proficiency in this vernacular, comparable to proficiencies in the scientific (numeracy) and humanities (literacy) disciplines. Design possesses its unique set of knowledge, methods of comprehension, and investigative approaches to this knowledge (Cross, 1982, pp. 221–222).

Davis (1998) claims that design strategies explicitly cater to a variety of learning styles and intelligences, thereby involving students from diverse backgrounds and abilities in their educational processes (p. 7). Consisting of several key steps, the standard design process, as described by Davis et al. (1997), is a systematic approach that starts with problem identification, followed by research and analysis, criteria setting for solutions, generation and prototyping of alternatives, selection and implementation of the best solution, and finally, outcome evaluation (Quintana et al., 2004). Crucial to this process is the research phase before prototyping, where designers study past solutions to similar problems to inform and enhance their new designs, ensuring innovation and effectiveness (Zhang et al., 2020). The design process emphasizes the development of various solutions through an iterative cycle of testing and refining (Davis, 1998). Both modeling and diagramming are highly valued alongside quantitative methods for their ability to elucidate the connections between concepts, catering to the diverse learning styles of students by leveraging their unique strengths (Rosa, 2016). Analysis contributes to this process, but the primary objective is the synthesis of a solution that meets the comprehensive set of performance criteria defined by the problem at hand (Davis et al., 1997; Fortus et al,. 2004).

When interpreting and evaluating this model in the context of language learning and teaching, it can be seen as a systematic approach to curriculum development, lesson planning, and educational innovation. Here's how each step could be applied:

- Identify and Define Problems: This is where the process begins. Designers identify the problems that need to be solved or the needs that should be addressed. This involves recognizing challenges in language education, such as students struggling with grammar or pronunciation. It's important to clearly define these issues to target them effectively.
- Gather and Analyze Information: Once the problem is defined, relevant information is collected and analyzed. This step could involve researching language acquisition theories, gathering data on student performance, and considering feedback from learners and educators. Understanding the context and the learners' needs is crucial.
- Set Success Criteria: Before solutions are generated, it's important to know what success looks like. Determining what success looks like in language learning—such as fluency levels, test scores, or the ability to communicate in real-life situations—helps in aiming for clear and measurable goals.
- Generate Solutions and Prototypes: With the problem, information, and criteria set, designers brainstorm possible solutions. Developing new teaching methods, materials, or technologies to address the defined problems. For example, creating interactive exercises for grammar practice or using software to improve pronunciation.
- Evaluate and Select Solutions: Testing the effectiveness of the developed solutions in the context of language learning. This can include pilot studies, experimental classes, or controlled trials to select the most effective methods.
- Implement Choices: Once the best solutions are selected, they are implemented in the language learning environment. This could mean integrating a new app into the classroom, adopting a new textbook, or training teachers in a new pedagogical approach.
- Evaluate Outcomes: Finally, assessing the impact of these changes on language learning outcomes. Are students improving? Are the goals being met? This evaluation can lead to further refinement of the solutions.

In a nutshell, the standard design process emphasizes an iterative and evidence-based approach to improving language education. Each step feeds into the next, creating a continuous loop of improvement. This model ensures that language teaching practices are constantly evolving and adapting to better meet the needs of learners.

Backwards Thinking in DBL

Similar to the instructional strategies which are all constructivist learning models, another DBL process was developed by Nelson (2006), known as Backwards Thinking®. The Doreen Nelson Method of Design-Based Learning offers a dynamic evolution, integrating these core principles with a unique emphasis on constructive student engagement and practical application in educational settings (Nelson, 2006). Doreen Nelson's Design-Based Learning, known as Backwards Thinking®, employs a unique 6½-step approach (Rosa, 2016). This methodology starts with the selection of the theme or concept to be taught, progresses to problem identification, introduces a novel design challenge (accounting for the half step), establishes the criteria, allows for student construction, integrates guided lessons, and culminates in student revisions of their creations (Ale, 2020). Contrary to the conventional DBL process that prioritizes research and information gathering before the design and construction phases, Nelson's strategy inverts this sequence, prompting students to engage in research post-construction (Fisher, 2018). This reversal facilitates a deeper level of cognitive engagement by challenging students to create without mimicking pre-existing structures like bridges, skyscrapers, or vehicles (Davis et al., 1997). The construction phase is followed by state standards-aligned lessons, after which students refine their designs (Nhem, 2015). This iterative process enables them to integrate their newfound understanding of real-world mechanics with their original designs, enhancing both their creativity and critical thinking skills (Na, 2019). Distinguishing the DBL approach in such a way is crucial as it mandates students to first construct and design the manipulative, compelling them to devise unique solutions to issues prior to examining historical and modern perspectives (Nelson, 1984).

Application of the Design Process in Language Learning

The DBL process by Nelson (2006) presents a cyclic, design-based model for language learning and teaching. Here's an interpretation and evaluation of the model in that context:

- Identify a theme or concept that underlies the curriculum: This is a crucial step in language learning, where educators identify core themes or concepts around which to center the learning experience. This could involve grammatical structures, vocabulary themes, cultural concepts, or language functions.
- Identify a problem from the curriculum: Educators pinpoint a specific problem or challenge related to the theme or concept that students need to overcome. In language learning, this could be a common linguistic mistake, a difficult grammatical concept, or a cultural misunderstanding.
- Turn the problem into a "never-before-seen" design challenge: This step involves creativity, where the problem is recontextualized into a unique challenge that students must solve. In language teaching, this could translate into a task-based

activity where students must use the language to solve a real-world problem, thereby making learning more engaging and practical.

- Set criteria for assessment, using the standards and content of the curriculum: Establishing clear criteria for assessment aligned with curriculum standards ensures that students understand the expectations and learning goals. In language teaching, this could involve rubrics that assess communicative competence, accuracy, fluency, and the ability to perform specific language functions.
- Let students "give it a try": Students engage with the challenge, applying what they know and experimenting with the language in a practical setting. This active engagement is essential for language acquisition as it provides real-life context and purpose for language use.
- Teach traditional guided lessons: This represents a more structured approach to language instruction, where specific language rules, vocabulary, and concepts are taught in a traditional classroom setting. This ensures that students receive explicit instruction on language forms and usage.
- Students revise designs: After receiving guided instruction and attempting the challenge, students reflect on their work and make revisions. In language learning, this could involve peer review, teacher feedback, and self-assessment, promoting an iterative learning process where students continuously improve their language skills.

The DBL process encourages active learning through real-world challenges, which can increase student motivation and engagement. The cyclic nature allows for continuous improvement and iterative learning, which is conducive to language acquisition. Incorporates both student-centered activities and teacher-guided instruction, providing a balanced approach. Assessment criteria linked to curriculum standards ensure that learning objectives are met. Overall, this model promotes a dynamic, interactive, and reflective approach to language learning that can be very effective if implemented with careful consideration of students' needs and abilities.

The Interaction of Mind and Hand Model

Dynamic Nature of the Design Process

The exploration of the design process might not fully encompass its dynamic essence. Typically, exploration as an endeavor is evaluated via a static research document, with the process being quantified by its outcomes. Educational institutions often transform dynamic activities into a sequence of static outcomes (Davis et al., 1997). Kimbell et al. (1991) aimed to find a suitable representation to illustrate the vibrant nature of the design process and created the Interaction of Mind and Hand model.

In the model, the interacting design loop illustrates the design process's divergent and cyclical characteristics (Davis et al., 1997). Kimbell et al. (1991) present a model that explains the interaction between the mind and hand, elucidating the reasons and methodologies behind students' choices in utilizing the design process, focusing on the rationale behind actions rather than the actions themselves (Looijenga et al., 2015, 2016). Additionally, the model demonstrates that design activities enhance the fluency of thought operations, namely the ability to think in both images and words, in a manner that purely reflective activities cannot achieve (Davis et al., 1997). Kimbell et al. (1991) articulate this by defining student engagement in the design process as "thought in action," thereby questioning the traditional separation between thought and action prevalent in numerous educational curricula. By prioritizing the interactive process in DBL, the emphasis shifts from the products themselves to the support they offer in this process (McCormick & Davidson, 1996). This approach places greater emphasis on the cognitive and decision-making activities that arise from interacting with these products, thereby allocating more balanced consideration to the "why" and "how" pupils chose to do things rather than "what" they chose to do (Todd, 1999). The learner visualizes and constructs an internal representation of knowledge, then juxtaposes this internal understanding with the external realities (Kimbell & Stables, 2007). Thus, the ideas and intentions of the pupils are deemed equally significant as the eventual products derived from them. Often, students are under the impression that they have devised a comprehensive solution right from the beginning of a task and are keen to materialize that concept into its final embodiment.

Cognitive and Physical Interactions in Learning

The "Interaction of Mind and Hand" model by Kimbell et al. (1991) illustrates a process where thinking and physical action (doing) are interconnected in the development of ideas and solutions. The interpretation and evaluation of this model within the context of language learning and teaching are as follows:

- Start with initial (hazy) idea

  Thinking: An idea forms in the mind, which is often vague or incomplete.
  Doing: The idea is expressed outwardly by talking to others, making drawings, sketches, notes, or graphs to represent the thought.

- Explore new idea, enhance the idea

  Thinking: The initial idea is developed through contemplation and enhancement in the mind.
  Doing: Creation of a rough model to represent the idea or prediction, transforming the thought into a tangible form.

- Clarify and validate idea

  Thinking: The idea is further refined and clarified, and attempts are made to validate it internally.
  Doing: A prototype is made based on the refined idea and tested in various situations to assess its viability.

- Critically appraise idea

  Thinking: The idea is critically evaluated to assess its strengths and weaknesses. Doing: This stage implies a continuation of the testing and refining process, possibly leading to new prototypes.

- Potential for more developed thinking

  Thinking: The critical appraisal feeds back into the thought process, suggesting potential for more sophisticated or advanced thinking about the idea.

- Potential for more developed solutions

  Doing: As a result of the advanced thinking and iterative process, there is potential for more developed, refined, and effective solutions.

In language teaching, the "Interaction of Mind and Hand" model underscores the importance of an active, learner-centered approach where language is not just absorbed passively but constructed through interaction, experimentation, and reflection. It aligns well with CT-Driven DBLL methodologies, which emphasize the use of language as a tool for making meaning and achieving communication goals.

Conclusion

Integration of CALL, DBL, and CT in Language Education

Upon exploring the topic from a historical perspective, the evolution of language education in the context of the digital age is reflected upon. In this sense, the historical progression of technology, particularly computers, within foreign language education forms a critical backdrop to the discussion. Beginning with the earliest language laboratories and moving towards today's advanced, interactive online platforms, the incorporation of ICT into language teaching has undergone a significant transformation. The term CALL, an acronym for Computer-Assisted or Computer-Aided Language Learning, effectively captures this evolution, representing the method of learning languages in various contexts with the aid of computer technologies. This wide-ranging definition underscores the integration of any kind of computer technology into a language learning setting, indicating a substantial departure from conventional to digital teaching methodologies. Amidst this shift towards digitalization, three distinct viewpoints have significantly shaped the convergence of language learning theories and general learning theories with ICT: the structural, the cognitive/constructivist, and the sociocognitive perspectives. Every point of view has contributed to the understanding of how technological advancements can be used to support language acquisition. This includes the drills and systematic repetition of the structural approach, the interesting, learner-centered activities emphasized by the ideas of cognitive/constructivists, and the dynamic, communicative practices promoted by the sociocognitive approaches. In this light, the integration of DBL and CT in language education can be seen as a contemporary application of socio-cognitive theory.

Summary of Key Points

This section explores the integration of CT with CALL through DBL practices, emphasizing the transformative potential in language education.

Key points:

- The historical evolution of CALL illustrates the shift from early language labs to advanced interactive online platforms.
- Three major perspectives—structural, cognitive/constructivist, and sociocognitive—have shaped the integration of technology in language learning.
- Integrating CT with DBL represents a contemporary application of socio-cognitive theory, fostering dynamic and interactive learning experiences.
- CT-driven DBL activities promote the analysis of language patterns, sentence construction and deconstruction, and logical reasoning.
- Misconceptions about the complexity of teaching CT are challenged, demonstrating its integration into language learning without extensive technical knowledge.
- The standard design process and principles of backward thinking DBL ensure alignment with educational objectives and deep engagement.
- Practical applications of CT and DBL in language learning enhance interaction and enjoyment in the learning process.
- The integration of CALL, DBL, and CT prepares learners for 21st-century linguistic demands and equips them with critical thinking and problem-solving skills.
- Continued innovation and collaboration among educators, researchers, and developers are essential to creating immersive and intellectually engaging language learning environments.

The integration of computational thinking, design-based learning, and computer-assisted language learning represents a significant advancement in language education. By merging these approaches, educators can create engaging, effective, and interactive learning environments that prepare students for the linguistic and cognitive demands of the digital age. This synergy not only enhances language proficiency but also fosters critical thinking and problem-solving skills, ensuring learners are well-equipped to navigate and contribute to an increasingly interconnected world. Continued innovation and collaboration in this field are essential for realizing the full potential of these educational methodologies.

Implications for Future Practice and Research

The integration of CT with CALL through DBL offers a transformative potential for language education, emphasizing problem-solving, pattern recognition, and algorithmic thinking. Future research should focus on:

- Developing and evaluating CT-driven DBL activities to enhance language learning outcomes.
- Investigating the long-term impacts of CT integration on language proficiency and digital literacy.
- Exploring the scalability of CT and DBL methodologies in diverse educational settings.
- Identifying best practices for teacher training and curriculum development to effectively implement these approaches.
- Examining the role of technology in facilitating collaborative and interactive language learning environments. For practice, educators should:
  - Incorporate CT principles into language lessons to foster analytical and problem-solving skills.
  - Utilize DBL frameworks to design engaging and contextually meaningful language learning activities.
  - Leverage technological tools to create immersive and interactive language learning experiences.
  - Provide continuous professional development to teachers on integrating CT and CALL in their teaching practices.
  - Engage students in real-world tasks that combine language learning with computational thinking to prepare them for future challenges.

References

Ackermann, E. (2001). Piaget's constructivism, Papert's constructionism: What's the difference. *Future of Learning Group Publication, 5*(3), 438. http://www.sylviastipich.com/wp-content/uploads/2015/04/Coursera-Piaget-_-Papert.pdf

Ale, M. (2020). *Framework for improving virtual learning environments (VLES) using pedagogical* [Unpublished Master's Thesis, University of Birmingham].

Alkaromah, A. H., Fauziati, E., & Asib, A. (2020). Students' perception on the Information and Communication Technology (ICT) tools in English language teaching. *ELS Journal on Interdisciplinary Studies in Humanities, 3*(1), 122–129. https://doi.org/10.34050/els-jish.v3i1.9522

Alonso-Prieto, V. (2019). *Computational thinking to boost motivation of students with learning difficulties in foreign language learning* [Final Degree Assignment, Universidad de Valladolid]. https://uvadoc.uva.es/bitstream/handle/10324/39642/TFG-G3966.pdf?sequence=5&isAllowed=y

Amazian, M., Soudane, J. A., Rhanoui, M., & Amazian, K. (2022). Proposal for a transformation project in higher education to reinforce foreign languages using the design thinking method.

In L. Daadaoui et al. (Eds.), *Proceedings of the Erasmus scientific days: Vol. 5* (pp. 290–303). Springer. https://doi.org/10.2991/978-2-38746-036-7_27
Apedoe, X. S., Reynolds, B., Ellefson, M. R., & Schunn, C. D. (2008). Bringing engineering design into high school science classrooms: The heating/cooling unit. *Journal of Science Education and Technology, 17*, 454–465. https://doi.org/10.1007/s10956-008-9114-6
Armstrong, E., & Ferguson, A. (2010). Language, meaning, context, and functional communication. *Aphasiology, 24*(4), 480–496. https://doi.org/10.1080/02687030902775157
Barr, D., Harrison, J., & Conery, L. (2011). Computational thinking: A digital age skill for everyone. *Learning & Leading with Technology, 38*(6), 20–23. https://files.eric.ed.gov/fulltext/EJ918910.pdf
Barr, J. D., & Gillespie, J. H. (2003). Creating a computer-based language learning environment. *ReCALL, 15*(1), 68–78. https://doi.org/10.1017/S0958344003000612
Beatty, K. (2013). *Teaching & researching: Computer-assisted language learning*. Routledge.
Bergroth-Koskinen, U. M., & Seppala, R. (2012). Teacher-researchers exploring design-based research to develop learning designs in higher education language teaching. *Apples-Journal of Applied Language Studies*, 6(2), 95–112. http://apples.jyu.fi/
Cameron, K. (Ed.). (1989). *Computer assisted language learning: Program structure and principles*. Intellect Books.
Cheung, Y. L., Chu, Y. N. J., & Jang, H. (2021). Impact of a socio-cognitive approach to teaching English language writing on primary school students' compositions. *Iranian Journal of Language Teaching Research, 9*(1), 1–22. https://doi.org/10.30466/ijltr.2021.120973
Chiu, W. Y., Liu, G. Z., Barrett, N. E., Liaw, M. L., Hwang, G. J., & Lin, C. C. (2023). Needs analysis-based design principles for constructing a context-aware English learning system. *Computer Assisted Language Learning, 36*(1–2), 176–204. https://doi.org/10.1080/09588221.2021.1905666
Chomsky, N. (1965). *Aspects of the theory of syntax*. MIT Press.
Cleminson, T., & Cowie, N. (2021). Using design thinking as an approach to creative and communicative engagement in the English as a Foreign Language (EFL) classroom. *Journal of University Teaching & Learning Practice, 18*(4), 1–20. https://doi.org/10.53761/1.18.4.7
Clifford, R. (1987). The status of computer-assisted language instruction. *CALICO Journal, 4*(4), 9–16. http://www.jstor.org/stable/24147523
Cohen, L. M. (1999). *Section III: Philosophical perspectives in education, Part 2*. http://oregonstate.edu/instruct/ed416/PP2.html
Conley, D. T. (2014). *The Common Core State Standards: Insight into their development and purpose*. Council of Chief State School Officers. https://csaa.wested.org/wp-content/uploads/2019/11/2014_CCSS_Insight_Into_Development.pdf
Costa, S., Gomes, A., & Pessoa, T. (2016). Using scratch to teach and learn English as a foreign language in elementary school. *International Journal of Education and Learning Systems, 1*, 207–213. https://iaras.org/iaras/filedownloads/ijels/2016/002-0027.pdf
Crites, K., & Rye, E. (2020). Innovating language curriculum design through design thinking: A case study of a blended learning course at a Colombian university. *System, 94*, 102334. https://doi.org/10.1016/j.system.2020.102334
Cross, N. (1982). Designerly ways of knowing. *Design Studies, 3*(4), 221–227. https://doi.org/10.1016/0142-694X(82)90040-0
Csizmadia, A., Standl, B., & Waite, J. (2019). Integrating the constructionist learning theory with computational thinking classroom activities. *Informatics in Education, 18*(1), 41–67. https://doi.org/10.15388/infedu.2019.03
Curzon, P., & McOwan, P. W. (2017). *The power of computational thinking: Games, magic and puzzles to help you become a computational thinker*. World scientific.
Davies, G., & Fitzpatrick, T. (2003). *The impact of information and communication technologies on the teaching of foreign languages and the role of teachers of foreign languages*. The Directorate General of Education and Culture. Retrieved September 18, 2004, from http://www.icceurope.com/verweise.htm

Davies, G. & Higgins, J. (1982). *Computers, language and language learning.* CILT Publications.
Davis, M. (1998). Making a case for design-based learning. *Arts Education Policy Review, 100*(2), 7–15. https://doi.org/10.1080/10632919809599450
Davis, M., Hawley, P., McMullan, B., & Spilka, G. (1997). *Design as a catalyst for learning.* Association for Supervision and Curriculum Development.
Echevarria, J., Vogt, M., & Short, D. (2008). *Making content comprehensible for English learners: The SIOP model.* Allyn & Bacon.
Egbert, J. (2005). CALL essentials. In *Principles and practice in CALL classrooms. TESOL.*
Ellis, D. G. (1999). *From language to communication.* Routledge.
Estes, C. (2004). Promoting student-centered learning in experiential education. *Journal of Experiential Education, 27*(2), 141–160. https://doi.org/10.1177/105382590402700203
Fillmore, L. W. (1991). Second-language learning in children: A model of language learning in social context. In E. Bialystok (Ed.), *Language processing in bilingual children* (pp. 49–69). Cambridge University Press.
Fisher, L. A. (2018). *Wonder city project: Teaching second grade curriculum with the application of the Doreen Nelson methodology of design-based learning.* California State Polytechnic University
Fortus, D., Dershimer, R. C., Krajcik, J., Marx, R. W., & Mamlok-Naaman, R. (2004). Design-based science and student learning. *Journal of Research in Science Teaching, 41*(10), 1081–1110. https://doi.org/10.1002/tea.20040
Garrett, N. (2009). Computer-assisted language learning trends and issues revisited: Integrating innovation. *The Modern Language Journal, 93*, 719–740. https://doi.org/10.1111/j.1540-4781.2009.00969.x
Gilmore, A. (2007). Authentic materials and authenticity in foreign language learning. *Language Teaching, 40*(2), 97–118. https://doi.org/10.1017/S0261444807004144
Gómez-Puente, S. M., Van Eijck, M., & Jochems, W. (2013). A sampled literature review of design-based learning approaches: A search for key characteristics. *International Journal of Technology and Design Education, 23*, 717–732. https://doi.org/10.1007/s10798-012-9212-x
Grover, S., & Pea, R. (2013). Computational thinking in K–12: A review of the state of the field. *Educational Researcher, 42*(1), 38–43. https://doi.org/10.3102/0013189X12463051
Gulruh, N., Dildora, S., Bunyod, B., & Fazliddin, Y. (2020). Interactive learning environment in the English language teaching with the help of computer technology. *Solid State Technology, 63*(4), 619–623.
Gunawardena, C. N., Chen, Y., Flor, N., & Sánchez, D. (2023). Deep learning models for analyzing social construction of knowledge online. *Online Learning, 27*(4).
Halliday, M. A. K. (1971). Language in a social perspective. *Educational Review, 23*(3), 165–188.
Healey, D. (2016). Language learning and technology: Past, present and future. In F. Farr & L. Murray (Eds.), *The Routledge handbook of language learning and technology* (pp. 38–55). Routledge.
Hsu, T. C., Chang, C., Wu, L. K., & Looi, C. K. (2022). Effects of a pair programming educational robot-based approach on students' interdisciplinary learning of computational thinking and language learning. *Frontiers in Psychology, 13*, 888215.
Hsu, T. C., Chang, S.-C., & Hung, Y.-T. (2018). How to learn and how to teach computational thinking: Suggestions based on a review of the literature. *Computers & Education, 126*, 296–310. https://doi.org/10.1016/j.compedu.2018.07.004
Hsu, T. C., & Liang, Y. S. (2021). Simultaneously improving computational thinking and foreign language learning: Interdisciplinary media with plugged and unplugged approaches. *Journal of Educational Computing Research, 59*(6), 1184–1207. https://doi.org/10.1177/0735633121992480
Hung, H. T. (2017). Design-based research: Redesign of an English language course using a flipped classroom approach. *TESOL Quarterly, 51*(1), 180–192.
Jacob, S. R., Parker, M. C., & Warschauer, M. (2022). Integration of computational thinking into English language arts. In A. Ottenbreit-Leftwich, & A. Yadav (Eds.), *Computational thinking*

in PreK-5: Empirical evidence for integration and future directions (pp. 55–63). https://doi.org/10.1145/3507951.3519288

Jacob, S., Nguyen, H., Tofel-Grehl, C., Richardson, D., & Warschauer, M. (2018). Teaching computational thinking to English learners. *NYS TESOL Journal, 5*(2), 12–24.

Jonassen, D., & Land, S. (2012). *Theoretical foundations of learning environments*. Routledge.

Kafai, Y. B., & Resnick, M. (1996). *Constructionism in practice: Designing, thinking, and learning in a digital world*. Routledge.

Kern, R. (2006). Perspectives on technology in learning and teaching languages. *TESOL Quarterly, 40*(1), 183–210. https://doi.org/10.2307/40264516

Kern, R., & Warschauer, M. (2000). Theory and practice of networked-based language teaching. In M. Warschauer & R. Kern (Eds.), *Network-based language teaching: Concepts and practice* (pp. 1–19). Cambridge University Press.

Khatib, M., & Shakouri, N. (2013). On situating the stance of socio-cognitive approach to language acquisition. *Theory & Practice in Language Studies, 3*(9), 1590–1595.

Kimbell, R., & Stables, K. (2007). *Researching design learning: Issues and findings from two decades of research and development*. Springer.

Kimbell, R., Stables, K., Wheeler, T., Wosniak, A., & Kelly, V. (1991). *The assessment of performance in design and technology*. School Examinations and Assessment Council.

Kirwan, C., Costello, E., & Donlon, E. (2018). *Computational thinking and online learning: A systematic literature review*. Academic Conferences Ltd.

Kramsch, C., & Thorne S. (2003). Foreign language learning as a global communicative practice. In D. Block, & D, Cameron (Eds.). *Language learning and teaching in the age of globalization* (pp. 83–100). Routledge.

Krashen, S. D. (1981). *Second language acquisition and second language learning*. Pergamon.

Kwan, Y. W., & Wong, A. F. (2015). Effects of the constructivist learning environment on students' critical thinking ability: Cognitive and motivational variables as mediators. *International Journal of Educational Research, 70*, 68–79. https://doi.org/10.1016/j.ijer.2015.02.006

Lee, L. (2007). Fostering L2 oral communication through constructivist interaction in desktop videoconferencing. *Foreign Language Annals, 40*, 635–649. https://doi.org/10.1111/j.1944-9720.2007.tb02885.x

Levy, M. (1997). *Computer-assisted language learning: Context and conceptualisation*. Oxford University Press.

Levy, M., & Hubbard, P. (2005). Why call CALL "CALL"? *Computer Assisted Language Learning, 18*(3), 143–149.

Li, Q. (2008). Knowledge building in an online environment: A design-based research study. *Journal of Educational Technology Systems, 37*(2), 195–216. https://doi.org/10.2190/ET.37.2.f

Looijenga, A., Klapwijk, R., & de Vries, M. J. (2015). The effect of iteration on the design performance of primary school children. *International Journal of Technology and Design Education, 25*, 1–23. https://doi.org/10.1007/s10798-014-9271-2

Looijenga, A., Klapwijk, R., & De Vries, M. J. (2016). Groundwork: Preparing an effective basis for communication and shared learning in design and technology education. *Design and Technology Education, 21*(3), 41–50. https://openjournals.ljmu.ac.uk/DATE/article/view/1579

Lu, J. J., & Fletcher, G. H. L. (2009). Thinking about computational thinking. *ACM SIGCSE Bulletin, 41*(1), 260–264. https://doi.org/10.1145/1539024.1508959

McCormick, R., & Davidson, M. (1996). Problem solving and the tyranny of product outcomes. *Journal of Design and Technology Education, 1*(3), 230–241.

Mehalik, M. M., Doppelt, Y., & Schuun, C. D. (2008). Middle-school science through design-based learning versus scripted inquiry: Better overall science concept learning and equity gap reduction. *Journal of Engineering Education, 97*(1), 71–85. https://doi.org/10.1002/j.2168-9830.2008.tb00955.x

Moreno, R., & Kilpatrick, J. (2018). Student perceptions of self-efficacy in the foreign language classroom: A design-based research study. *EDeR. Educational Design Research, 2*(1), 1–26. https://doi.org/10.15460/eder.2.1.1214

Mubarok, H., Lin, C. J., & Hwang, G. J. (2023). A virtual reality-based collaborative argument mapping approach in the EFL classroom. *Interactive Learning Environments*, 1–19. https://doi.org/10.1080/10494820.2023.2207197

Na, S. (2019). *Starter city: Implementing the Doreen Nelson design-based learning methodology and the common core state standards* [Master's Thesis, California State Polytechnic University].

Nelson, D. (1984). *Transformations: Process and theory*. Center for City Building Educational Programs.

Nelson, D. (2006). *The 6½ steps of backwards thinking™*. http://www.cpp.edu/~dnelson/methodology/6.5steps.html

Nhem, L. (2015). *Teachers' application of design-based learning principles toward creating a common core aligned curriculum* [Doctoral dissertation, California State Polytechnic University].

Nieto, S. (2001). *Language, culture, and teaching: Critical perspectives*. Routledge.

Pande, M., & Bharathi, S. V. (2020). Theoretical foundations of design thinking–A constructivism learning approach to design thinking. *Thinking Skills and Creativity, 36*, 100637.

Papert, S. (1980). *Mindstorms: Children, computers, and powerful ideas*. BasicBooks, Inc.

Papert, S. (1990). *A critique of technocentrism in thinking about the school of the future. Epistemology and learning memo #2. September 1990*. MIT. Retrieved August 29, 2015 from www.papert.org/articles/ACritiqueofTechnocentrism.html

Papert, S. (1993). *The children's machine: Rethinking school in the age of the computer*. Basic Books.

Parmaxi, A., & Zaphiris, P. (2016). Computer-mediated communication in computer-assisted language learning: Implications for culture-centered design. *Universal Access in the Information Society, 15*, 169–177.

Perrenet, J. C., Bouhuijs, P. A., & Smits, J. G. (2000). The suitability of problem-based learning for engineering education: Theory and practice. *Teaching in Higher Education, 5*(3), 345–358.

Piaget, J. (1954). *The construction of reality in the child*. Ballantine Books.

Pica, T. (1996). The essential role of negotiation in the communicative classroom. *Jalt Journal, 18*(2), 241–268.

Quintana, C., Reiser, B. J., Davis, E. A., Krajcik, J., Fretz, E., Golan Duncan, R., Kyza, E., Edelson, D., & Soloway, E. (2004). A scaffolding design framework for software to support science inquiry. *Journal of the Learning Sciences, 13*(3), 337–386. https://doi.org/10.1207/s15327809jls1303_4

Reeves, T. C., & McKenney, S. (2013). Computer assisted language learning and design-based research: Increased complexity for sure, enhanced impact perhaps. In J. C. Rodríguez, & C. Pardo-Ballester (Eds.), *Design-based research in CALL* (pp. 9–21). Computer Assisted Language Instruction Consortium (CALICO).

Rodríguez, A. J. T. (2022). ICTS and computational thinking to foster language skills: A proposal based on Scratch. *Didácticas Específicas, 27*, 8–30.

Rosa, R. (2016). *Design-based learning: a methodology for teaching and assessing creativity* [Doctoral Dissertation, California State Polytechnic University].

Rottenhofer, M. (2022). *Modeling in the foreign language classroom: A hands-on approach to foster computational thinking skills* [Doctoral Dissertation, Johannes Kepler University].

Rottenhofer, M., Kuka, L., Leitner, S., & Sabitzer, B. (2022). Using computational thinking to facilitate language learning: A survey of students' strategy use in Austrian secondary schools. *IAFOR Journal of Education, 10*(2), 51–70.

Rottenhofer, M., Sabitzer, B., & Rankin, T. (2021). Developing computational thinking skills through modeling in language lessons. *Open Education Studies, 3*(1), 17–25.

Roy, D., & Brine, J. (2013). Design thinking in EFL context: Studying the potential for language teaching and learning. *International Journal of Design Education, 6*(2), 1–21.

Saad, A., & Zainudin, S. (2022). A review of project-based learning (PBL) and computational thinking (CT) in teaching and learning. *Learning and Motivation, 78*, 101802.

Sabitzer, B. (2012). Spreadsheets for language learning: Creative ideas for informatics and foreign language lessons. In *CSEDU 2012—Proceedings of the 4th international conference on computer supported education.*
Sabitzer, B. (2013). Informatics meets foreign languages COOL ideas for a cross-curricular cooperation. *Computers in Human Behavior, 29*(2), 424–432.
Sabitzer, B., Demarle-Meusel, H., & Jarnig, M. (2018). Computational thinking through modeling in language lessons. In *2018 IEEE global engineering education conference (EDUCON)* (pp. 1913–1919). IEEE.
Shetzer, H., & Warschauer, M. (2000). An electronic literacy approach to network-based language teaching. *Network-Based Language Teaching: Concepts and Practice, 3*(5), 171–185.
Stahl, G., Koschmann, T., & Suthers, D. (2014). Computer-supported collaborative learning. In R. K. Sawyer (Ed.), *The Cambridge handbook of the learning sciences* (pp. 479–500). Cambridge University Press.
Stapa, S. H. (2007). Socio-cognitive theory in second language learning: The use of on-line forum among adult distance learners. *International Journal of Learning, 14*(7), 137–142.
Stewart, A. E., Vrzakova, H., Sun, C., Yonehiro, J., Stone, C. A., Duran, N. D., Shute, V., & D'Mello, S. K. (2019). I say, you say, we say: Using spoken language to model socio-cognitive processes during computer-supported collaborative problem solving. *Proceedings of the ACM on Human-Computer Interaction, 3*(CSCW), 1–19.
Swain, M., & Watanabe, Y. (2013). Languaging: Collaborative dialogue as a source of second language learning. *The Encyclopedia of Applied Linguistics*, 3218–3225.
Thomas, M. (2013). Task-based language teaching and CALL. *Contemporary Computer-Assisted Language Learning*, 341–358.
Todd, R. (1999). Design and technology yields a new paradigm for elementary schooling. *Journal of Technology Studies, 25*(2), 26–33.
Van den Branden, K., Bygate, M., & Norris, J. (2009). *Task-based language teaching: A reader.* John Benjamins.
Vigmo, S. (2010). *New spaces for language learning: A study of student interaction in media production in English* [Doctoral dissertation, University of Gothenburg].
Vogel, F., Wecker, C., Kollar, I., & Fischer, F. (2017). Socio-cognitive scaffolding with computer-supported collaboration scripts: A meta-analysis. *Educational Psychology Review, 29*, 477–511.
Warschauer, M. (1996). Computer assisted language learning: An introduction. In S. Fotos (Ed.), *Multimedia language teaching* (pp. 3–20). Logos International.
Warschauer, M., & Kern, R. (Eds.). (2000). *Network-based language teaching: Concepts and practice.* Cambridge University Press.
Wing, J. M. (2006). Computational thinking. *Communications of the ACM, 49*(3), 33–35.
Wolz, U., Stone, M., Pearson, K., Pulimood, S. M., & Switzer, M. (2011). Computational thinking and expository writing in the middle school. *ACM Transactions on Computing Education (TOCE), 11*(2), 1–22.
Wu, T. T., Silitonga, L. M., & Murti, A. T. (2024). Enhancing English writing and higher-order thinking skills through computational thinking. *Computers & Education*, 105012.
Youjun, T., & Xiaomei, M. (2022). Computational thinking: A mediation tool and higher-order thinking for linking EFL grammar knowledge with competency. *Thinking Skills and Creativity, 46*, 101143.
Zhang, F., Markopoulos, P., & Bekker, T. (2020). Children's emotions in design-based learning: A systematic review. *Journal of Science Education and Technology, 29*, 459–481.
Zhao, J. H., Chen, Z. W., & Yang, Q. F. (2024). I do and I understand: A virtual reality-supported collaborative design-assessing activity for EFL students. *System, 121*, 103213.

Chapter 7
CT Integration by Design-Based Learning Perspective—Implementation Example of Foreign Language Training

Ali Duran

Abstract This chapter investigates how to apply design-based learning (DBL) to the integration of computational thinking (CT) into language learning. In this sense, the main premise of this chapter is to examine the potential of integrating CT and DBL approaches for enhancing language learning outcomes. A selection of case studies extracted from existing literature is explored to illustrate the successful application of CT principles within DBL contexts. The theoretical underpinnings and the design framework of a CT-driven design-based language learning (DBLL) project are then outlined, emphasizing the importance of a solid foundation in both CT and language learning pedagogies. The execution phase is detailed, showcasing the application of CT processes such as problem decomposition, pattern recognition, and algorithmic thinking in language learning tasks. Furthermore, the evaluation methodologies employed to assess the effectiveness of the project, including qualitative feedback and quantitative data, are discussed, followed by reflections on the outcomes and the learning process. The necessity of professional development for educators to effectively integrate CT in language learning is emphasized, including training in both CT concepts and their application within DBLL. The development and availability of curricular resources that support the integration of CT in language learning, facilitating a smoother transition for educators and learners alike, are highlighted. Various formative assessment techniques that enable the ongoing evaluation of students' understanding and application of CT principles in language learning are explored. Summative assessment methods are discussed, focusing on their role in measuring the overall effectiveness of CT integration in achieving language learning objectives. Finally, the chapter concludes with a discussion on the potential future directions for CT-driven DBLL, emphasizing the need for further research, the development of innovative teaching strategies, and the exploration of new technologies to support this pedagogical approach.

Keywords CT-driven language education · Teacher preparation and support · Professional development · Curricular resources · Assessment strategies

A. Duran (✉)
Department of Foreign Language Education, Amasya University, Amasya, Türkiye
e-mail: aliduran@amasya.edu.tr

M. Saritepeci and H. Yildiz Durak (eds.), *Integrating Computational Thinking Through Design-Based Learning*, https://doi.org/10.1007/978-981-96-0853-9_7

Introduction

Learning, a complex phenomenon, cannot be isolated from its context as educational innovations are introduced into real-life contexts, offering the possibility to study change or the effects of interventions in those settings (Bell, 2004; Bergroth-Koskinen & Seppala, 2012). In response to the rapid development of technology and the emergence of new professions, twenty-first-century education is shifting towards a competence-oriented approach, emphasizing skills like CT that will be beneficial in the future (Pan et al., 2023). Teachers of the twenty-first century, facing fundamental educational challenges, must adapt to changes at an ever-increasing pace to prepare today's pupils for the modern job market and educate them for an uncertain future (Wing, 2006).

The swift rise in the use of social or Web 2.0 technologies has resulted in an extensive array of research across different educational settings, showcasing the diverse ways these technologies enhance language teaching and learning. (e.g., Bennett et al., 2012; Jalkanen & Vaarala, 2013; Liu et al., 2016). Nevertheless, inquiries into the adoption of social technologies within the field of language education confronts with dual concerns related to their theoretical foundations and pedagogical coherence. Firstly, most studies is not theoretically grounded (Wang & Vasquez, 2012), whereas the integration of Web 2.0 technologies in the domain of language pedagogy demands better task-technology alignment (Parmaxi & Zaphiris, 2016). Social technologies are not inherently linked to a specific educational methodology. Therefore, educators and professionals must thoughtfully select and apply educational theories to inform the instructional choices guiding the implementation of these technologies (Ruschoff & Ritter, 2001).

As it is stated by Wing (2006), CT transcends the boundary of computer science, establishing itself as a universal skill and mindset relevant across all disciplines. As such, CT can be seamlessly integrated with a multitude of subjects, enriching them with its principles (Hsu et al., 2018). In fact, CT has been widely used in different subjects, including mathematics (Benakli et al., 2017), biology (Libeskind-Hadas & Bush, 2013), computer science (Grover et al., 2015), language (Evia et al., 2015) and programming (Bers et al., 2014). The growing global interest in CT underscores the necessity for more comprehensive investigations from diverse viewpoints (Weng & Wong, 2017). When it comes to language education itself, CT as a problem-solving process and modeling as a visualization tool can support the demanding processes that occur in foreign language learning (Newport, 1990). According to the literature, there have already been attempts to implement CT in language lessons (e.g. Barr & Stephenson, 2011; Lu & Fletcher, 2009). Despite its wide applicability, it appears that only a select few researchers have explored the promising intersection between CT and language learning. This oversight suggests a significant untapped area ripe for investigation, where the analytical frameworks of CT could potentially revolutionize traditional language education methodologies (Ge et al., 2015), so more inquiries needed on CT in language education (Bergroth-Koskinen & Seppala, 2012). Within the field of English language education, such facets as listening, speaking, reading and writing exist. The integration of CT merely acts as an initial exploration into its applicability in language learning. Therefore, further investigations are imperative to determine the feasibility of integrating CT into other dimensions of language education (Weng & Wong, 2017). Last but not least, despite the widespread adoption of computers and mobile devices in language education within schools and their popularity in research, the integration of CT principles among language educators remains relatively rare and is minimally explored (Sabitzer et al., 2018).

The integration of CT into foreign language education can significantly enhance learning outcomes by promoting systematic problem-solving skills and critical thinking (Hsu et al., 2022; Li et al., 2022). CT encourages learners to recognize patterns in language, apply algorithmic thinking to language learning strategies, and develop a deeper understanding of linguistic structures through decomposition and abstraction (Rottenhofer et al., 2021). These skills not only facilitate a more efficient acquisition of a new language but also empower learners to connect language learning with real-world applications, making the learning process more relevant and engaging (Bergroth-Koskinen & Seppala, 2012). There has been empirical evidence that enhancing students' CT skills positively impact a variety of cognitive and metacognitive abilities (Israel-Fishelson & Hershkovitz, 2022; Yadav et al., 2022). Research also suggests that the implementation of CT techniques has been shown to improve English language skills (Dijaya et al., 2017; Parsazadeh et al., 2020; Weng & Wong, 2017).

Learning strategies addressed within the scope of CT learning activities are classified as aesthetic experience (Farris & Sengupta, 2016), collaborative learning (Saad, 2020), concept-based learning (Boudah et al., 2000), critical computational literacy (Tsai et al., 2021), embodied learning (Pierson & Brady, 2020), game-based learning (Hooshyar et al., 2021), human–computer interaction teaching (McCoy & Ketterlin-Geller, 2004), problem solving system (Yildiz-Durak, 2020), problem-based learning (Kwon et al., 2021), project-based learning (Saad & Zainudin, 2022), scaffolding (Tikva & Tambouris, 2023), digital storytelling (Sarıtepeci & Durak, 2016), student-centered lecture (Broza et al., 2023), systematic computational strategies (Alajlan et al., 2023) and universal design for learning (Israel et al., 2020) by Hsu et al. (2018). More specifically, among the learning strategies listed here, especially algorithmic language learning strategies (Li, 2021), content-based teaching (Brinton et al., 1989), digital storytelling (Huang, 2023, Parsazadeh et al., 2020), game-based language learning (Chowdhury et al., 2024), interdisciplinary media with plugged and unplugged approaches (Hsu & Liang, 2021) and mobile assisted language learning (Fronza & Gallo, 2016) are frequently used strategies in language education. There is also another common learning strategy called design-based language learning (DBLL). Significantly, DBL emerges as a prominent strategy within the scope of CT and stands out as a common and effective approach in foreign language education, acting as a pivotal actor enhancing the educational landscape with its innovative and engaging methodologies (Roy & Brine, 2013; Roy et al., 2020; Vigmo, 2010). More specifically, EFL grammar, as explored by Youjun and Xiaomei (2022), alongside vocabulary development, as discussed by Tang and Ma (2024), deeply intertwines with cognitive processes. It's all about remembering knowledge, understanding ideas or concepts through pattern recognition, applying knowledge through abstraction, weaving connections through parallelization, assessing learning through decomposition, and crafting syntactic writing with the help of coding or data analysis as well as debugging, describing, and comparing (Jacob et al., 2022; Rottenhofer et al., 2022).

DBLL is an innovative educational strategy that merges language learning with design thinking principles (Cleminson & Cowie, 2021; Nazim & Mohammad, 2022). This approach encourages learners to engage in the creative process of designing projects, artifacts, or solutions while using the target language (El-Sakka, 2020). Through DBLL, students apply language in authentic, meaningful contexts, thereby promoting active learning, critical thinking, and the practical application of language skills in real-life problems (Crites & Rye, 2020). Integrating CT within DBLL frameworks further amplifies its effectiveness, enabling students to systematically approach problem-solving and project design with computational and linguistic tools (Ko & Lim, 2022).

Integrating CT into language learning and teaching through DBL offers a promising approach to address the pedagogical and theoretical challenges highlighted in the use of social or Web 2.0 technologies (Rosa, 2016). CT-Driven DBL Activities in language learning leverage the principles of systematic problem solving, algorithmic thinking, and the iterative design process inherent in CT to foster a deeper understanding of language concepts and enhance learners' ability to use language creatively and effectively (Bergroth-Koskinen & Seppala, 2012). This integration is a perfect fit as it aligns with the need for theoretical grounding and task-technology alignment in language learning contexts (Weng & Wong, 2017). Theoretical benefits of this approach include promoting a more structured and analytical mindset among learners, encouraging them to deconstruct language into its fundamental components, and explore patterns and rules in a manner similar to coding (Sabitzer et al., 2018). Practically, CT-Driven DBL activities facilitate authentic, real-world language use, allowing learners to engage in meaningful projects that require communication, collaboration, and critical thinking (Parmaxi et al., 2016). These activities also offer opportunities for personalized learning experiences and can be easily adapted to leverage social technologies, thus addressing the gap in task-technology alignment (Vigmo, 2010). CT integration through DBL not only enriches language learning with a robust theoretical framework but also enhances practical language skills, preparing learners for the digital age. In this chapter, the pedagogical frameworks that support this integration are outlined, the benefits and challenges associated with its implementation are discussed, and evidence demonstrating CT's impact on language education is provided.

Implementation Examples: Best Practices on CT-Driven DBLL

So far, the groundwork has been laid by intertwining CT and DBL with language learning, highlighting the role of technology in this fusion. The next aim is to analyze previous literature and demonstrate real-world applications in CT-driven DBLL.

Design-Based Research (DBR) in Language Learning

Higher Education: In their study, Bergroth-Koskinen and Seppala (2012) investigated the application of Design-Based Research (DBR) in the context of higher education language teaching. The purpose is to adapt and improve learning designs to better support students in becoming proficient communicators and self-directed learners. The methodology involves iterative cycles of designing, implementing, and refining learning interventions based on feedback and analysis, emphasizing real-world classroom settings. Findings highlight the potential of DBR to enhance learner agency and the effective communication of expertise, suggesting that well-designed learning environments can significantly impact students' ability to navigate and contribute to their fields of study effectively.

Multimedia Environments: The research by Hung (2011) significantly contributes to understanding by investigating the use of multimedia environments to support English language learning through design-based research (DBR). Focusing on reflective tasks enhanced by digital video technology, the research aimed to tailor instructional designs for effective language learning. Key findings indicate that multimedia tools, particularly video recordings and reflective practices, significantly aid students in understanding and improving their presentation skills. Reflective tasks, designed iteratively, fostered students' engagement and deepened their learning processes, highlighting the potential of integrating multimedia technology in language education to enhance both theoretical understanding and practical application.

Development of Online Reading Materials: Through their study, Pardo-Ballester and Rodriguez (2010) provide compelling evidence on the development of online reading materials for Spanish learners through design-based research (DBR), examining interface design and learner perceptions. Key findings include the importance of learner involvement in material design, the effectiveness of multimedia glosses for vocabulary learning, and the iterative nature of developing educational materials. The research underscores the potential of DBR in creating engaging and effective language learning tools, highlighting the dynamic between content, interface design, and learner interaction.

Systematic Reviews: Ozverir et al. (2021) performed a systematic quantitative literature review of design-based research (DBR) in technology-enhanced language learning (TELL) from 2013 to 2018. It aims to synthesize past studies on the integration of technology and educational theories in language learning, identifying trends and offering insights for future research. The study highlights the use of DBR to address educational problems, develop effective interventions, and enhance theoretical understanding within TELL. It underscores the importance of DBR in creating practical and theoretical advancements in the field.

Social Technologies in Language Learning: Parmaxi et al. (2016) embark on a comprehensive study on the use of social technologies in language learning, structured around a design-based research (DBR) approach. It explores how collaborative construction of shareable artifacts using social technologies can enhance language learning, guided by constructionist learning theories. The research unfolds over three iterative cycles, each refining the educational intervention based on insights from the previous one. Key findings include the importance of extending classroom activities into real-life contexts, the synergistic use of multiple social technologies, and the promotion of meaningful classroom dynamics that foster collaboration and engagement. Instructional design elements derived from the study emphasize the integration of real-life experiences, collaborative artifact creation, and the leveraging of social technologies to support learning processes. The research highlights the dynamic interplay between social technologies, task design, and classroom culture in promoting effective learning experiences.

Integration of Computational Thinking (CT) in Language Learning

Mobile-Assisted Language Learning: The work of Fronza and Gallo (2016) examines the integration of CT into mobile-assisted language learning, specifically for German. It proposes an innovative methodology where mobile app exercises designed to enhance grammatical understanding also require the application of CT. The approach aims to improve language learning by making learners more aware of linguistic structures and evaluating their problem-solving skills. Through sample applications and a proof-of-concept system, the paper illustrates the potential benefits of integrating CT into language learning, highlighting the dual focus on grammatical competence and computational skills development.

Educational Robots and Board Games: Hsu and Liang (2021) explore the effectiveness of integrating CT and English learning through both plugged (using educational robots) and unplugged (using traditional board games) approaches in elementary education. The findings indicate that both methods significantly improved students' English vocabulary and sentence proficiency as well as their CT skills. The plugged approach was particularly effective in reducing students' foreign language learning anxiety and enhancing their critical thinking abilities, whereas the unplugged approach improved cooperation among students. The research suggests the potential of incorporating educational technology to simultaneously advance language learning and CT in young learners.

Pair Programming with Educational Robots: In another study on DBLL, Hsu et al. (2022) investigate the impact of a pair programming educational robot-based approach on students' interdisciplinary learning, specifically focusing on CT and language learning. It explores how this method affects students' CT skills, language learning progress, and anxiety. Utilizing educational robots in pair programming settings, the research examines the differences in learning outcomes between students studying Chinese as a Second Language (CSL) and those studying English as a Foreign Language (EFL). The study finds that this approach can significantly enhance both CT and language learning skills, with differences observed in the levels of learning anxiety and the specific competencies developed between the two groups.

Modeling in Language Lessons: The study by Rottenhofer et al. (2021) explores the development of CT skills through modeling in language lessons. It highlights the potential of using computer science models, particularly in foreign language teaching, to enhance CT without the need for technology. The research focuses on graphic models, specifically from computer science, to foster CT skills and presents several models useful in language teaching. Key findings suggest modeling facilitates understanding complex grammatical structures, enhances vocabulary learning, and improves problem-solving skills in language learning contexts.

Scratch Programming for English Learning: The study by Weng and Wong (2017) explore integrating Computational Thinking (CT) into English dialogue learning using the graphical programming tool Scratch. It aimed to investigate the classroom effect of this integration and how it could enhance English education with programming elements. Through class observations, programming projects, and interviews with nine primary students, the study found that introducing CT into English learning through Scratch motivated students and suggested further exploration into connecting CT with language education. The study also proposed making computers interact more autonomously with students and incorporating game design in programming environments to foster English learning.

Design Thinking and Language Learning

Legal English and German for IT Students: In a study by Opincāne and Laganovska (2023), it was demonstrated that the application of the design thinking approach in the development of foreign language courses, specifically targeting Bachelor level law students learning legal English and Information Technology students studying German as a second foreign language. It highlights the effectiveness of Design Thinking in enhancing communicative competencies and motivation among learners. The research incorporates methods such as survey data analysis and qualitative content analysis, with findings suggesting that this approach fosters real-life problem-solving, collaboration, and active engagement in professional and real-life scenarios.

The study concludes that Design Thinking is particularly beneficial for learners with higher levels of foreign language proficiency.

Web Design Analysis for EFL: Roy and Brine (2013) explore integrating design thinking into English as a Foreign Language (EFL) education, and examined how web design analysis can foster higher-order thinking and enhance language learning through content-based instruction. The research involved computer science students analyzing web designs to improve language skills, focusing on interface, navigation, and content analysis. Findings indicate that structured design activities, combined with regular feedback, can significantly contribute to developing systematic thinking and language production in EFL learners.

Modeling and CT Tools: Sabitzer et al. (2018) focus on the integration of CT through modeling in language lessons. The study explores the potential of CT tools, like modeling, to enhance language learning in primary and secondary education. It emphasizes the dual benefits of teaching computational concepts and improving language skills, such as text comprehension and vocabulary acquisition. The paper presents creative ways to introduce CT in language classes and shares experiences and results from teacher observations, interviews, and feedback from students and teachers. Key findings suggest that modeling can support language learning in various ways, helping to structure and visualize grammatical rules and improve problem-solving skills. The research indicates that CT can be seamlessly integrated into language lessons, making it accessible to students without a computer science background.

Digital Media and Language Learning

Film Production: Vigmo (2010) examines how digital media, specifically film production, integrates into language learning. It explores the role of digital media in creating new learning environments and opportunities for language practice outside traditional classrooms. The methodology is design-based research focusing on upper secondary students producing a film in English, capturing their interactions, storyboarding, and the final production through videos, observations, and interviews. Findings reveal that digital media facilitates new forms of language production and interaction, encouraging improvisation, code-switching, and the application of media literacy. The research highlights the emergence of hybrid practices blending media literacy with language learning, suggesting that these practices can significantly enhance foreign language education by leveraging students' existing media repertoires and promoting more engaging, relevant, and personalized learning experiences.

Designing a CT-Driven DBLL Project: An Example

Here, an example of a CT-Driven DBLL project based on non-linear digital storytelling is presented here. This is because non-linear digital storytelling in CT-Driven DBLL enhances CT skills (Cecere et al, 2023), encourages interactive learning (Letonsaari, 2021), fosters creative and critical thinking (Smith & Burrow, 2023), improves language proficiency (Özüdoğru & Çakır, 2021), reflects real-world communication (Winkler et al., 2006), and utilizes technology effectively (Soleimani et al., 2019). This CT-Driven DBLL project covers the inception of the project idea, the planning and design phases. Additionally, it details the execution of the project, including how CT was integrated with DBLL objectives, the tools and resources utilized, and the assessment of student outcomes. This personal example aims to provide a practical perspective on applying CT-DBLL principles in real-world educational settings.

Foundation and Framework

Contextual Overview

Context and Participants: This project is designed for secondary school students learning a foreign language (e.g., English, Spanish, French). It is intended for a classroom setting that accommodates collaborative group work, with access to computers or tablets and internet connectivity. The ideal group size is 3–4 students, allowing for diverse roles and collaboration while ensuring each member can significantly contribute.

Settings and Demographics: The activity is suitable for both urban and suburban schools with basic to moderate technological resources. It assumes students have a foundational understanding of the target language and basic digital literacy skills. The diversity in linguistic backgrounds and prior exposure to technology among students can enrich the project outcomes through varied perspectives.

Educational Objectives and Competencies

- To enhance students' proficiency in the target language through the creation of digital stories.
- To develop CT skills by engaging in the design and development process of a non-linear digital storytelling application.

Linguistic Competence

- Improve vocabulary, grammar, and syntax understanding in the target language.
- Develop storytelling skills, including narrative structure and character development.
- Enhance listening and reading comprehension through peer story exchange.

CT Skill Development

- Decomposition: Break down the story creation process into manageable parts (e.g., scripting, designing characters, creating scenes).
- Pattern Recognition: Identify common narrative structures and themes in stories.
- Abstraction: Generalize from specific story examples to understand story elements that can be programmatically represented (e.g., characters, settings, plot sequences).
- Algorithm Design: Create a sequence of steps (algorithm) for users to construct their stories within the app, considering user interface design and user experience.

Execution

Designing the Learning Experience

- Lesson Sequence: The project will be structured over several weeks, with each phase building upon the previous one. Initial lessons will focus on introducing CT and digital storytelling concepts, followed by hands-on activities that guide students through the design and development of their storytelling app.
- Problem Identification and Decomposition: Begin by presenting students with the challenge of creating a digital storytelling app. Encourage them to identify the main components of this problem, such as story creation, app functionality, and user interface design. Break down these components into smaller, manageable tasks.
- Defining the Language Task: The core language task involves crafting a narrative in the target language. This includes developing a plot, characters, setting, and dialogue. Emphasize the use of rich, descriptive language and various grammatical structures to tell engaging stories.
- Identifying Sub-tasks: Break the language task into sub-tasks, such as vocabulary building around specific themes, practicing dialogue creation, and structuring narratives. Assign these as both individual and group activities to scaffold learning and ensure comprehensive language exposure.

Engaging with CT through DBLL

- Pattern Recognition and Abstraction: Guide students in identifying patterns in story structure and linguistic features. Use these patterns to abstract general principles of storytelling and language use that can be applied to their digital stories.
- Analyzing Linguistic Structures: Encourage students to analyze the linguistic structures common to storytelling in the target language, such as past tense verbs for narration, conditional phrases for hypothetical situations, and connectors to sequence events.
- Creating Language Models: Assist students in creating language models that can be used in their app, such as templates for character dialogues or descriptive paragraphs about settings, incorporating key vocabulary and grammatical structures.

- Algorithm Design and Implementation: Introduce the concept of algorithms by having students outline the step-by-step process a user might follow to create a story using their app. This includes selecting story elements, inputting text, and navigating through the app.
- Developing Language Strategies: Through the design and development process, students will naturally develop strategies for language learning, such as using context clues to understand new vocabulary or experimenting with syntax to convey different meanings.
- Applying CT Concepts (CT Integration): Explicitly connect CT concepts to language learning tasks. For example, use decomposition when breaking down the process of story creation or algorithm design when planning language learning activities.
- Real-world Context: Encourage students to create stories that reflect real-world experiences or cultural aspects related to the target language, enhancing the relevance and engagement of the learning experience.
- Collaboration and Reflection: Foster collaboration by having students work in teams and reflect on their learning process. Reflection activities might include discussing challenges faced, strategies used, and how CT concepts helped in language learning and app development.

Evaluation and Reflection

Assessing Language Performance

Evaluation of language performance will be based on both the process and the product. Assess students' ability to use the target language effectively in their storytelling, including grammar accuracy, vocabulary richness, and overall fluency. Use rubrics that consider creativity, language use, and adherence to storytelling conventions.

Reflecting on CT Application

- Reflect on the application of CT concepts throughout the project. This could involve self-assessment checklists or group discussions about how students applied decomposition, pattern recognition, abstraction, and algorithm design to both the language tasks and the app development process.
- Encourage students to consider how these CT skills can be transferred to other learning areas or real-world problems.

Some Fundamentals of CT Integration in DBLL

Teacher Preparation and Support

The role of educators is pivotal in integrating CT into DBL. Their preparation and ongoing support are crucial for the successful adoption and execution of this pedagogical approach.

Professional Development

Educators should have access to professional development opportunities that enhance their understanding of both CT and DBL concepts (Carlson & Gadio, 2002). This is because the initiation into integrating CT within DBL begins with comprehensive professional development for educators (Hestness et al., 2018). This is not just about familiarizing them with the core principles of CT and DBL but also about bridging these concepts with practical application in language learning contexts (Cabrera et al., 2023). Professional development programs should be designed to:

- Enhance Understanding of CT and DBL: Educators need to build a solid foundation in both CT and DBL. This includes understanding the theoretical underpinnings, the pedagogical implications, and the potential for innovation within language learning (Rosa, 2016).
- Strategies for Integration: Training should provide educators with a repertoire of strategies to seamlessly integrate CT into language learning. This could involve problem-solving activities, CT exercises that complement language learning objectives, and methods to encourage creative thinking and innovation among students (Weng & Wong, 2017).
- Technology Tools Facilitation: Given the digital nature of CT, educators must be adept at using technology tools that facilitate the integration of CT into language learning. This encompasses a range of digital tools, platforms, and resources that can support DBLL activities, including programming environments, digital storytelling tools, and collaborative online platforms (Jocius et al., 2021).
- Design and Implementation of CT-Driven Activities: The ultimate goal of professional development is to equip educators with the skills and knowledge to design and implement effective CT-driven language learning activities. This includes the ability to create engaging, challenging, and relevant learning experiences that leverage CT skills to enhance language acquisition and use (Rehmat et al, 2020).

Curricular Resources

In addition to professional development, teachers need access to high-quality curricular resources that support the integration of CT into language learning (Mills et al.,

2021). This includes lesson plans, project templates, and examples of successful CT-driven DBL activities. Curricular resources should also offer guidance on adapting materials to accommodate diverse learning needs and language proficiency levels, ensuring that all students can benefit from this approach (Waterman et al., 2020). These resources serve as a foundation for teachers to develop and refine their teaching strategies, ensuring that the integration of CT into language learning is both effective and seamless (Kwon et al., 2023). To this end, curricular resources should encompass a broad range of materials and tools designed to support educators at every step of the teaching and learning process (Yang et al., 2021). Curricular resources should include detailed lesson plans and project templates that are specifically tailored for integrating CT into language learning (Belmar, 2022). These plans and templates should:

- Outline clear learning objectives that align with both language learning goals and CT competencies (Prottsman, 2022).
- Provide step-by-step instructions for conducting CT-driven DBL activities in the language learning classroom (Voon et al., 2022).
- Suggest methods for incorporating digital tools and platforms that facilitate CT and language learning (Hsu & Liang, 2021).
- Include assessment strategies to evaluate students' language development and their understanding of CT concepts (Rottenhofer et al., 2021).

Assessment Strategies

Formative Assessment Techniques

Formative assessment techniques are employed throughout the learning process to monitor student progress, providing ongoing feedback that helps learners improve their language and CT skills (Hadad et al., 2020). Techniques include observational assessments, where educators observe students during task completion to assess their problem-solving approaches and linguistic application in real-time (Bonner et al., 2021). Peer review sessions, where students critique and provide feedback on each other's work, encourage reflection on both language use and CT strategies (Bergroth-Koskinen & Seppala, 2012). Additionally, reflective journals, in which students document their learning processes, challenges, and strategies for overcoming obstacles, offer insights into the integration of CT principles in language tasks (Poulakis & Politis, 2021).

Summative Assessment Techniques

Summative assessment techniques are utilized at the conclusion of a learning module or project to evaluate overall student achievement in both language learning and CT (Román-González et al., 2019). These include project-based assessments, where students demonstrate their ability to apply language skills and CT principles in a final project, and standardized tests that measure CT skills in a controlled environment (Polat et al., 2021). Portfolios, compiling a selection of students' work throughout the course, provide a comprehensive overview of their growth and the application of CT in language learning tasks. These techniques offer concrete evidence of the efficacy of CT and DBL integration in enhancing linguistic competence and CT (Fields et al., 2021).

Conclusion

The integration of CT into language learning has emerged as a pivotal development in educational pedagogy, reflecting the evolving landscape of twenty-first-century skills and competencies. CT not only enhances the traditional learning process by introducing an analytical mindset to linguistic studies but also prepares learners for the digital era, where problem-solving and critical thinking are invaluable. The inclusion of CT in language education transcends the mere acquisition of vocabulary and grammar; it fosters a deeper understanding of language structures, encourages innovative thinking, and facilitates the application of language skills in real-world scenarios. The empirical evidence presented underlines the versatility and effectiveness of CT when integrated with DBL methodologies, particularly in facilitating meaningful and engaging language learning experiences. Through CT-driven DBL activities, learners are equipped to analyze and deconstruct language patterns, apply logical reasoning, and engage in hands-on problem-solving, mirroring the processes inherent in computational disciplines. This approach not only deepens linguistic comprehension but also democratizes access to CT education, proving that it does not necessitate extensive technical knowledge or resources. Moreover, the studies reviewed emphasize the significant role of educators in this integration process. Their preparation, support, and access to curricular resources and professional development opportunities are crucial for the successful implementation of CT-driven language learning activities. Assessment strategies, both formative and summative, are key to monitoring and evaluating student progress, ensuring that learners achieve desired outcomes in both language proficiency and CT skills.

Looking forward, the integration of CT into language learning through DBLL presents a landscape ripe for exploration and innovation. Future research should aim to expand the understanding of how CT principles can be more effectively woven into various language learning contexts, including online and blended learning environments, to cater to the diverse needs of learners across different age groups and proficiency levels. The development of scalable and adaptable CT-driven DBL frameworks that can be customized to fit different linguistic and cultural contexts is essential. These frameworks should also consider the rapid advancements in technology, incorporating new digital tools and platforms that can enhance the language learning experience. Additionally, there is a pressing need for more comprehensive teacher training programs that equip educators with the skills to seamlessly integrate CT into their language teaching practices. Interdisciplinary research can provide new insights into the cognitive processes underlying language learning and CT, offering innovative strategies for curriculum development. In conclusion, the integration of CT in language learning through DBLL represents a forward-thinking approach that not only enhances educational outcomes but also equips learners with the skills necessary to thrive in a digitally interconnected world.

Summary of Key Points

This section investigates the synergy between computational thinking (CT) and design-based learning (DBL) in language education, highlighting the innovative methodologies and their impacts on enhancing language learning experiences.

Key points:

- Integrating Computational Thinking (CT) within Design-Based Learning (DBL) frameworks significantly enhances language learning outcomes.
- This integration promotes systematic problem-solving skills, critical thinking, and language proficiency.
- Utilizing digital storytelling, educational robots, and other technology tools enriches the learning experience.
- Both plugged (technology-enhanced) and unplugged (traditional) methods are effective in fostering CT and language skills.
- Educators play a pivotal role in the successful integration of CT into language learning.
- Comprehensive professional development is essential to equip teachers with the necessary skills and strategies.
- High-quality curricular resources, including lesson plans and project templates, are crucial for effective CT-driven DBL activities.
- Resources should be adaptable to diverse learning needs and proficiency levels.
- Formative assessment techniques such as observational assessments, peer reviews, and reflective journals monitor ongoing progress.
- Summative assessments, including project-based evaluations and portfolios, provide a comprehensive measure of language and CT skill development.
- CT-driven DBL activities improve vocabulary, grammar, narrative skills, and overall language fluency.
- These activities facilitate real-world language application, making learning more relevant and engaging.
- Further research is needed to explore the integration of CT in various language learning contexts, including online and blended environments.
- Development of scalable, adaptable CT-driven DBL frameworks and comprehensive teacher training programs is essential.

By embedding CT principles within DBL, this chapter demonstrates a forward-thinking approach that not only enhances language learning but also prepares students for a digitally interconnected world.

Implications for Future Practice and Research

1. **Research on CT Integration**: Future research should investigate the long-term impacts of CT integration on language proficiency and digital literacy. Studies should explore the scalability of CT and DBL methodologies across diverse educational settings.
2. **Development of Innovative Teaching Strategies**: Educators should develop and implement innovative teaching strategies that leverage CT principles to enhance language learning. This includes creating engaging and contextually meaningful learning experiences.
3. **Teacher Training Programs**: Comprehensive teacher training programs are essential for the successful integration of CT into language learning. These programs should focus on building educators' understanding of CT and DBL, providing practical strategies, and familiarizing them with relevant technological tools.
4. **Curriculum and Resource Development**: There is a need for the development of scalable and adaptable curricular resources that support CT integration. These resources should cater to different linguistic and cultural contexts and incorporate advancements in technology.
5. **Assessment and Evaluation**: Implementing effective formative and summative assessment methods is crucial for monitoring student progress and evaluating the success of CT-driven DBL activities. These assessments should measure both language proficiency and CT skills.
6. **Interdisciplinary Collaboration**: Interdisciplinary research can offer new insights into the cognitive processes underlying language learning and CT. Collaborative efforts between language educators, computer scientists, and educational researchers can lead to the development of innovative curriculum designs and teaching methodologies.
7. **Exploration of New Technologies**: The rapid advancement of technology presents opportunities for incorporating new digital tools and platforms into language learning. Future research should explore the use of emerging technologies, such as virtual reality and artificial intelligence, to create immersive and interactive learning environments.

Appendix: Lesson Plan

Lesson Plan: CT-Driven DBLL Project Based on Non-Linear Digital Storytelling

Title: Enhancing Language Proficiency and Computational Thinking through Non-Linear Digital Storytelling.

Duration: 8 Weeks.

Grade Level: Secondary School Students.

Subjects: Foreign Language (e.g., English, Spanish, French), Computational Thinking.

Overview: This lesson plan outlines a project-based learning experience where students will create non-linear digital stories to enhance their language proficiency and develop computational thinking (CT) skills. The project integrates design-based learning (DBL) principles and digital tools to foster creative and critical thinking, improve language skills, and reflect real-world communication.

Objectives:

- Enhance proficiency in the target language through digital storytelling.
- Develop CT skills by designing and developing a non-linear digital storytelling application.
- Improve vocabulary, grammar, and narrative skills in the target language.
- Foster collaboration, creativity, and problem-solving abilities.

Materials:

- Computers or tablets with internet connectivity
- Digital storytelling tools (e.g., Scratch, Twine, Storybird)
- Rubrics for assessment
- Reflection journals (Table A.1).

Contextual Overview:

- **Context and Participants**: Designed for secondary school students learning a foreign language (e.g., English, Spanish, French). The project accommodates collaborative group work, with access to computers or tablets and internet connectivity. Ideal group size: 3–4 students.
- **Settings and Demographics**: Suitable for both urban and suburban schools with basic to moderate technological resources. Assumes students have a foundational understanding of the target language and basic digital literacy skills.

Educational Objectives and Competencies:

- **Linguistic Competence**:
  - Improve vocabulary, grammar, and syntax understanding in the target language.

Table A.1 Detailed lesson plan for CT-driven DBLL project based on non-linear digital storytelling

Week	Objectives	Activities	Assessment
1	Understand the basics of CT and digital storytelling	– Introduction to CT concepts: decomposition, pattern recognition, abstraction, algorithm design – Introduction to non-linear digital storytelling – Explore examples of digital stories – Group discussion on project goals and expectations	– Participation in discussions – Understanding of basic concepts
2	Begin planning the digital storytelling project and develop story ideas	– Brainstorm story ideas in groups – Identify main project components: story creation, app functionality, user interface design – Break down the project into manageable tasks (decomposition) – Develop a project timeline and assign roles	– Initial story outlines – Project plans
3	Define the language task and build relevant vocabulary	– Develop plot, characters, setting, and dialogue for the stories – Vocabulary building exercises around specific themes – Practice creating dialogues and descriptive paragraphs	– Vocabulary tests – Draft of the narrative
4	Design and create the elements of the digital story	– Create character dialogues, settings, and plot sequences – Use digital tools to design and implement these elements – Apply CT concepts (e.g., pattern recognition, abstraction) to story design	– Story elements created – Application of CT concepts
5	Develop the non-linear digital storytelling application	– Introduce the concept of algorithms and user interface design – Outline the step-by-step process for users to create their stories in the app – Begin programming the digital story using chosen tools	– Progress in app development – Clarity of user interface design
6	Test the digital storytelling app and refine it based on feedback	– Peer review sessions to test each other's apps – Gather feedback and identify areas for improvement – Refine and debug the app based on peer feedback	– Peer review feedback – Refined app versions

(continued)

Table A.1 (continued)

Week	Objectives	Activities	Assessment
7	Finalize the digital stories and prepare for presentations	– Finalize all elements of the digital story and app – Prepare presentations to showcase the digital stories – Rehearse presentations and receive feedback from peers	– Completed digital stories – Presentation preparation
8	Present digital stories and reflect on the learning process	– Present digital stories to the class – Reflect on the application of CT concepts and language learning – Group discussions and self-assessment on project experiences – Encourage students to consider how CT skills can be applied to other areas	– Presentation quality – Reflection journals – Self-assessment checklists

- Develop storytelling skills, including narrative structure and character development.
- Enhance listening and reading comprehension through peer story exchange.

• **CT Skill Development**:

- **Decomposition**: Break down the story creation process into manageable parts (e.g., scripting, designing characters, creating scenes).
- **Pattern Recognition**: Identify common narrative structures and themes in stories.
- **Abstraction**: Generalize from specific story examples to understand story elements that can be programmatically represented (e.g., characters, settings, plot sequences).
- **Algorithm Design**: Create a sequence of steps (algorithm) for users to construct their stories within the app, considering user interface design and user experience

Evaluation and Reflection:

• **Assessing Language Performance**:

- Use rubrics to assess grammar accuracy, vocabulary richness, storytelling creativity, and adherence to conventions.

• **Reflecting on CT Application**:

- Self-assessment checklists or group discussions about how students applied decomposition, pattern recognition, abstraction, and algorithm design to both the language tasks and the app development process.
- Encourage students to consider how these CT skills can be transferred to other learning areas or real-world problems.

References

Alajlan, H., Alebaikan, R., & Almassaad, A. (2023). Computational thinking in K–12 computer education: Appropriate pedagogy. *Technology, Pedagogy and Education, 32*(3), 337–349. https://doi.org/10.1080/1475939X.2023.2184857

Barr, V., & Stephenson, C. (2011). Bringing computational thinking to K-12: What is involved and what is the role of the computer science education community? *ACM Inroads, 2*(1), 48–54. https://doi.org/10.1145/1929887.1929905

Bell, P. (2004). On the theoretical breadth of design-based research in education. *Educational Psychologist, 39*(4), 243–253. https://doi.org/10.1207/s15326985ep3904_6

Belmar, H. (2022). Review on the teaching of programming and computational thinking in the world. *Frontiers in Computer Science, 4*, 997222. https://doi.org/10.3389/fcomp.2022.997222

Benakli, N., Kostadinov, B., Satyanarayana, A., & Singh, S. (2017). Introducing computational thinking through hands-on projects using R with applications to calculus, probability and data analysis. *International Journal of Mathematical Education in Science and Technology, 48*(3), 393–427. https://doi.org/10.1080/0020739X.2016.1254296

Bennett, S., Bishop, A., Dalgarno, B., Waycott, J., & Kennedy, G. (2012). Implementing Web 2.0 technologies in higher education: A collective case study. *Computers & education, 59*(2), 524–534. https://doi.org/10.1016/j.compedu.2011.12.022

Bergroth-Koskinen, U. M., & Seppala, R. (2012). Teacher-researchers exploring design-based research to develop learning designs in higher education language teaching. *Apples-Journal of Applied Language Studies, 6*(2), 95–112. http://apples.jyu.fi/

Bers, M. U., Flannery, L., Kazakoff, E. R., & Sullivan, A. (2014). Computational thinking and tinkering: Exploration of an early childhood robotics curriculum. *Computers & Education, 72*, 145–157. https://doi.org/10.1016/j.compedu.2013.10.020

Bonner, S., Chen, P., Jones, K., & Milonovich, B. (2021). Formative assessment of computational thinking: Cognitive and metacognitive processes. *Applied Measurement in Education, 34*(1), 27–45. https://doi.org/10.1080/08957347.2020.1835912

Boudah, D. J., Lenz, B. K., Bulgren, J. A., Schumaker, J. B., & Deshler, D. D. (2000). Don't water down! Enhance content learning through the unit organizer routine. *Teaching Exceptional Children, 32*(3), 48–56. https://doi.org/10.1177/004005990003200З

Brinton, D. M., Snow, M. A. & Wesche, M. B. (1989). *Content-based second language instruction.* Heinle & Heinle.

Broza, O., Biberman-Shalev, L., & Chamo, N. (2023). Start from scratch: Integrating computational thinking skills in teacher education program. *Thinking Skills and Creativity, 48*, 101285. https://doi.org/10.1016/j.tsc.2023.101285

Cabrera, L., Ketelhut, D. J., Mills, K., Killen, H., Coenraad, M., Byrne, V. L., & Plane, J. D. (2023). Designing a framework for teachers' integration of computational thinking into elementary science. *Journal of Research in Science Teaching*, 1–36. https://doi.org/10.1002/tea.21888

Carlson, S., & Gadio, C. T. (2002). Teacher professional development in the use of technology. In W. D. Haddad & A. Draxler (Eds.), *Technologies for education: Potentials, parameters, and prospects* (pp. 118–132). UNESCO.

Cecere, L., Colace, F., Lombardi, M., Lorusso, A., Santaniello, D., & Valentino, C. (2023, September). Cultural heritage enhancement through digital storytelling and context-aware recommender system. In *Proceedings of the 20th ınternational conference on content-based multimedia ındexing* (pp. 86–91). https://doi.org/10.1145/3617233.3617241

Chowdhury, M., Dixon, L. Q., Kuo, L. J., Donaldson, J. P., Eslami, Z., Viruru, R., & Luo, W. (2024). Digital game-based language learning for vocabulary development. *Computers and Education Open, 6*, 100160. https://doi.org/10.1016/j.caeo.2024.100160

Cleminson, T., & Cowie, N. (2021). Using design thinking as an approach to creative and communicative engagement in the English as a Foreign Language (EFL) classroom. *Journal of University Teaching & Learning Practice, 18*(4), 1–20. https://doi.org/10.53761/1.18.4.7

Crites, K., & Rye, E. (2020). Innovating language curriculum design through design thinking: A case study of a blended learning course at a Colombian university. *System, 94*, 102334. https://doi.org/10.1016/j.system.2020.102334

Dijaya, S., Wang, G., & Makbul, D. S. (2017). Innovation in English language teaching for EFL context: Students' perceptions toward writing story activity using computational thinking process. *Journal of Education and Practice, 8*(22), 72–78.

El Sakka, S. M. (2020). A design thinking–based strategy for enhancing EFL students' writing performance and their sustainability awareness. *Journal of Education, 80*(80), 15–42.

Evia, C., Sharp, M. R., & Pérez-Quiñones, M. A. (2015). Teaching structured authoring and DITA through rhetorical and computational thinking. *IEEE Transactions on Professional Communication, 58*(3), 328–343.

Farris, A. V., & Sengupta, P. (2016). Democratizing children's computation: Learning computational science as aesthetic experience. *Educational Theory, 66*(1–2), 279–296.

Fields, D., Lui, D., Kafai, Y., Jayathirtha, G., Walker, J., & Shaw, M. (2021). Communicating about computational thinking: Understanding affordances of portfolios for assessing high school students' computational thinking and participation practices. *Computer Science Education, 31*(2), 224–258. https://doi.org/10.1080/08993408.2020.1866933

Fronza, I., & Gallo, D. (2016). Towards mobile assisted language learning based on computational thinking. In: D. Chiu, I. Marenzi, U. Nanni, M. Spaniol, & M. Temperini (eds) *Advances in web-based learning—ICWL 2016. Lecture Notes in Computer Science* (Vol. 10013). Springer. https://doi.org/10.1007/978-3-319-47440-3_16

Ge, X., Ifenthaler, D., & Spector, J. (2015). Moving forward with STEAM education research. In: X. Ge, D. Ifenthaler, & J. M. Spector, (Eds.), *Emerging technologies for steam education. educational communications and technology: Issues and innovations*. Springer, Cham. https://doi.org/10.1007/978-3-319-02573-5_20

Grover, S., Pea, R., & Cooper, S. (2015). Designing for deeper learning in a blended computer science course for middle school students. *Computer Science Education, 25*(2), 199–237.

Hadad, R., Thomas, K., Kachovska, M., & Yin, Y. (2020). Practicing formative assessment for computational thinking in making environments. *Journal of Science Education and Technology, 29*, 162–173.

Hestness, E., Jass-Ketelhut, D., McGinnis, J. R., Plane, J., Razler, B., Mills, K., Cabrera, L., & Gonzalez, E. (2018). Computational thinking professional development for elementary science educators: Examining the design process. In E. Langran, & J. Borup (Eds.), *Proceedings of society for information technology & teacher education international conference* (pp. 1904–1912). Association for the Advancement of Computing in Education (AACE).

Hooshyar, D., Pedaste, M., Yang, Y., Malva, L., Hwang, G. J., Wang, M., & Lim, H., & Delev, D. (2021). From gaming to computational thinking: An adaptive educational computer game-based learning approach. *Journal of Educational Computing Research, 59*(3), 383–409.

Hsu, T. C., & Liang, Y. S. (2021). Simultaneously improving computational thinking and foreign language learning: Interdisciplinary media with plugged and unplugged approaches. *Journal of Educational Computing Research, 59*(6), 1184–1207.

Hsu, T. C., Chang, C., Wu, L. K., & Looi, C. K. (2022). Effects of a pair programming educational robot-based approach on students' interdisciplinary learning of computational thinking and language learning. *Frontiers in Psychology, 13*, 888215.

Hsu, T.-C., Chang, S.-C., & Hung, Y.-T. (2018). How to learn and how to teach computational thinking: Suggestions based on a review of the literature. *Computers & Education, 126*, 296–310. https://doi.org/10.1016/j.compedu.2018.07.004

Huang, H. T. D. (2023). Examining the effect of digital storytelling on English speaking proficiency, willingness to communicate, and group cohesion. *Tesol Quarterly, 57*(1), 242–269.

Hung, H. T. (2011). Design-based research: Designing a multimedia environment to support language learning. *Innovations in Education and Teaching International, 48*(2), 159–169.

Israel, M., Jeong, G., Ray, M., & Lash, T. (2020, February). Teaching elementary computer science through universal design for learning. In *Proceedings of the 51st ACM Technical Symposium on Computer Science Education* (pp. 1220–1226).

Israel-Fishelson, R., & Hershkovitz, A. (2022). Studying interrelations of computational thinking and creativity: A scoping review (2011–2020). *Computers & Education, 176*, 104353.

Jacob, S. R., Parker, M. C., & Warschauer, M. (2022). Integration of computational thinking into English language arts. In *Computational thinking in preK-5: Empirical evidence for integration and future directions* (pp. 55–63).

Jalkanen, J., & Vaarala, H. (2013). Digital texts for learning Finnish: Shared resources and emerging practices. *Language Learning & Technology, 17*(1), 107–124. 10125/24512.

Jocius, R., O'Byrne, W. I., Albert, J., Joshi, D., Robinson, R., & Andrews, A. (2021). Infusing computational thinking into STEM teaching. *Educational Technology & Society, 24*(4), 166–179.

Ko, E. G., & Lim, K. Y. (2022). Promoting English learning in secondary schools: Design-based research to develop a mobile application for collaborative learning. *The Asia-Pacific Education Researcher, 31*(3), 307–319.

Kwon, K., Jeon, M., Guo, M., Yan, G., Kim, J., Ottenbreit-Leftwich, A. T., & Brush, T. A. (2023). Computational thinking practices: Lessons learned from a problem-based curriculum in primary education. *Journal of Research on Technology in Education, 55*(4), 590–607.

Kwon, K., Ottenbreit-Leftwich, A. T., Brush, T. A., Jeon, M., & Yan, G. (2021). Integration of problem-based learning in elementary computer science education: Effects on computational thinking and attitudes. *Educational Technology Research and Development, 69*, 2761–2787.

Letonsaari, M. (2021). Nonlinear storytelling approach to developing computational thinking skills. *Helsinki Studies in Education.*

Li, W. (2021). English Teaching Design in Language Learning from Pascal's Error Checking Function. In *Application of Big Data, Blockchain, and Internet of Things for Education Informatization: First EAI International Conference, BigIoT-EDU 2021, Virtual Event, August 1–3, 2021, Proceedings, Part I 1* (pp. 140–148). Springer International Publishing.

Li, X., van Braak, J., Valcke, M., & Sang, G. (2022). A systematic literature review on integrating computational thinking into English language courses learning and teaching in K-12 education. In *EDULEARN22 proceedings* (pp. 4842–4850).

Libeskind-Hadas, R., & Bush, E. (2013). A first course in computing with applications to biology. *Briefings in Bioinformatics, 14*(5), 610–617.

Liu, C. C., Wang, P. C., & Tai, S. J. D. (2016). An analysis of student engagement patterns in language learning facilitated by Web 2.0 technologies. *ReCALL, 28*(2), 104–122.

Lu, J. J., & Fletcher, G. H. (2009, March). Thinking about computational thinking. In *Proceedings of the 40th ACM technical symposium on Computer science education* (pp. 260–264).

McCoy, J. D., & Ketterlin-Geller, L. R. (2004). Rethinking instructional delivery for diverse student populations: Serving all learners with concept-based instruction. *Intervention in School and Clinic, 40*(2), 88–95.

Mills, K., Coenraad, M., Ruiz, P., Burke, Q., & Weisgrau, J. (2021). *Computational thinking for an inclusive world: a resource for educators to learn and lead.* Digital Promise.

Nazim, M., & Mohammad, T. (2022). Implications of design thinking in an EFL classroom: Writing in context. *Theory and Practice in Language Studies, 12*(12), 2723–2730.

Newport, E. L. (1990). Maturational constraints on language learning. *Cognitive Science, 14*(1), 11–28.

Opincāne, M., & Laganovska, K. (2023, July). Applying of design thinking approach in the development of foreign language course. In *Society. Integration. Education. Proceedings of the international scientific conference* (Vol. 1, pp. 155–164). https://doi.org/10.17770/sie2023vol1.7102

Ozverir, I., Mokhtari, L., & Ozverir, A. (2021). A systematic analysis of design-based research in technology-enhanced language learning (TELL). *HAYEF: Journal of Education, 18*(3), 323–353.

Özüdoğru, G., & Çakır, H. (2021). Non-linear digital storytelling: Effect on technology utilization and writing self-efficacy. *Technology in Society, 67*, 101798.

Pan, Z., Cui, Y., Leighton, J. P., & Cutumisu, M. (2023). Insights into computational thinking from think-aloud interviews: A systematic review. *Applied Cognitive Psychology, 37*(1), 71–95.

Pardo-Ballester, C., & Rodriguez, J. C. (2010). Developing Spanish online readings using design-based research. *CALICO Journal, 27*(3), 540–553.

Parmaxi, A., & Zaphiris, P. (2016). Computer-mediated communication in computer-assisted language learning: Implications for culture-centered design. *Universal Access in the Information Society, 15*, 169–177.

Parmaxi, A., Zaphiris, P., & Ioannou, A. (2016). Enacting artifact-based activities for social technologies in language learning using a design-based research approach. *Computers in Human Behavior, 63*, 556–567.

Parsazadeh, N., Cheng, P.-Y., Wu, T.-T., & Huang, Y.-M. (2020). Integrating computational thinking concept ınto digital storytelling to ımprove learners' motivation and performance. *Journal of Educational Computing Research.* https://doi.org/10.1177/0735633120967315

Pierson, A. E., & Brady, C. E. (2020). Expanding opportunities for systems thinking, conceptual learning, and participation through embodied and computational modeling. *Systems, 8*(4), 48.

Polat, E., Hopcan, S., Kucuk, S., & Sisman, B. (2021). A comprehensive assessment of secondary school students' computational thinking skills. *British Journal of Educational Technology, 52*(5), 1965–1980.

Poulakis, E., & Politis, P. (2021). Computational thinking assessment: literature review. In *Research on e-learning and ICT in education: Technological, pedagogical and ınstructional perspectives* (pp. 111–128).

Prottsman, K. (2022). *Computational thinking meets student learning: Extending the ISTE standards.* International Society for Technology in Education.

Rehmat, A. P., Ehsan, H., & Cardella, M. E. (2020). Instructional strategies to promote computational thinking for young learners. *Journal of Digital Learning in Teacher Education, 36*(1), 46–62.

Román-González, M., Moreno-León, J., & Robles, G. (2019). Combining assessment tools for a comprehensive evaluation of computational thinking interventions. In *Computational thinking education* (pp. 79–98). Springer.

Rosa, R. (2016). *Design-based learning: a methodology for teaching and assessing creativity* [Doctoral Dissertation, California State Polytechnic University].

Rottenhofer, M., Kuka, L., Leitner, S., & Sabitzer, B. (2022). Using computational thinking to facilitate language learning: A survey of students' strategy use in Austrian secondary schools. *IAFOR Journal of Education, 10*(2), 51–70.

Rottenhofer, M., Sabitzer, B., & Rankin, T. (2021). Developing computational thinking skills through modeling in language lessons. *Open Education Studies, 3*(1), 17–25.

Roy, D., & Brine, J. (2013). Design thinking in EFL context: Studying the potential for language teaching and learning. *International Journal of Design Education, 6*(2), 1–21.

Roy, D., Kudry, P., & Naya, K. (2020). Analyzing the communicative efficacy of a soft CLIL focused project-based language learning EFL course on Smart homes. In *SHS web of conferences* (Vol. 77, p. 02002). EDP Sciences.

Ruschoff, B., & Ritter, M. (2001). Technology-Enhanced language learning: construction of knowledge and template-based learning in the foreign language classroom. *Computer Assisted Language Learning, 14*(3–4), 219–232.

Saad, A. (2020). Students' computational thinking skill through cooperative learning based on hands-on, inquiry-based, and student-centric learning approaches. *Universal Journal of Educational Research, 8*(1), 290–296.

Saad, A., & Zainudin, S. (2022). A review of project-based learning (PBL) and computational thinking (CT) in teaching and learning. *Learning and Motivation, 78*, 101802.

Sabitzer, B., Demarle-Meusel, H., & Jarnig, M. (2018). Computational thinking through modeling in language lessons. In *2018 IEEE global engineering education conference (EDUCON).* https://doi.org/10.1109/educon.2018.8363469

Sarıtepeci, M., & Durak, H. (2016). Digital storytelling effect on students' motivation in subject of information technologies fundamentals. *Journal of Research in Education and Teaching, 5*(Special Issue), 258–265.

Smith, S., & Burrow, L. E. (2023). Programming multimedia stories in Scratch to integrate computational thinking and writing with elementary students. *Journal of Mathematics Education, 9*(2), 119–131.

Soleimani, A., Herro, D., & Green, K. E. (2019). CyberPLAYce—A tangible, interactive learning tool fostering children's computational thinking through storytelling. *International Journal of Child-Computer Interaction, 20*, 9–23.

Tang, Y., & Ma, X. (2024). Enhancing EFL vocabulary acquisition through computational thinking. In: Zhao, F., & Miao, D. (Eds.), *AI-generated content. AIGC 2023. Communications in computer and ınformation science* (Vol. 1946). Springer. https://doi.org/10.1007/978-981-99-7587-7_6

Tikva, C., & Tambouris, E. (2023). The effect of scaffolding programming games and attitudes towards programming on the development of Computational Thinking. *Education and Information Technologies, 28*(6), 6845–6867.

Tsai, M. J., Liang, J. C., & Hsu, C. Y. (2021). The computational thinking scale for computer literacy education. *Journal of Educational Computing Research, 59*(4), 579–602.

Vigmo, S. (2010). *New spaces for language learning: A study of student interaction in media production in English* [Doctoral dissertation, University of Gothenburg].

Voon, X. P., Wong, S. L., Wong, L. H., Khambari, M. N. M., & Abdullah, S. I. S. S. (2022). Developing computational thinking competencies through constructivist argumentation learning: A problem-solving perspective. *International Journal of Information and Education Technology*.

Wang, S., & Vasquez, C. (2012). Web 2.0 and second language learning: what does the research tell us. *Calico, 29*(3), 412–430.

Waterman, K. P., Goldsmith, L., & Pasquale, M. (2020). Integrating computational thinking into elementary science curriculum: An examination of activities that support students' computational thinking in the service of disciplinary learning. *Journal of Science Education and Technology, 29*, 53–64.

Weng, X., & Wong, G. K. (2017, December). Integrating computational thinking into English dialogue learning through graphical programming tool. In *2017 IEEE 6th ınternational conference on teaching, assessment, and learning for engineering (TALE)* (pp. 320–325). IEEE.

Wing, J. M. (2006). Computational thinking. *Communications of the ACM, 49*(3), 33–35.

Winkler, T., Herczeg, M., & Goldmann, A. (2006, March). Why and what children learn while creating an interactive, non linear Mixed-Reality-Storytelling-Room. In *Society for ınformation technology & teacher education ınternational conference* (pp. 742–749). Association for the Advancement of Computing in Education (AACE).

Yadav, A., Ocak, C., & Oliver, A. (2022). Computational thinking and metacognition. *TechTrends, 66*(3), 405–411.

Yang, D., Baek, Y., Ching, Y. H., Swanson, S., Chittoori, B., & Wang, S. (2021). Infusing computational thinking in an integrated STEM curriculum: user reactions and lessons learned. *European Journal of STEM Education, 6*(1), 4.

Yildiz-Durak, H. (2020). The effects of using different tools in programming teaching of secondary school students on engagement, computational thinking and reflective thinking skills for problem solving. *Technology, Knowledge and Learning, 25*, 179–195.

Youjun, T., & Xiaomei, M. (2022). Computational thinking: A mediation tool and higher-order thinking for linking EFL grammar knowledge with competency. *Thinking Skills and Creativity, 46*, 101143.

Chapter 8
CT Integration by Design-Based Learning Perspective—Implementation Example of Digital Storytelling

Uğur Ferhat Ermiş and Hasan Celal Balıkçı

Abstract Computational thinking (CT) has received much attention in educational contexts in the recent years. One way to integrate Information and Communication Technology (ICT) is using design-based learning (DBL) that focuses on cyclical learning through designing, making, and refining responses to complex issues. One of the widely practiced DBL activities is digital storytelling (DS) that refers to the production and dissemination of multimedia stories using digital tools. This book chapter introduces DS as a strategy for integrating ICT into the curriculum. The chapter also outlines how the DS project is initiated, with a selection of a problem or theme, followed by the creation of story, and the use of digital tools to produce the story in the format of multimedia. The chapter further discusses the ICT principles that are part of the activity, such as problem decomposition, pattern recognition, abstraction and algorithm generation. In addition, it describes the benefits of using ICT through DS including the improvement of students' problem-solving skills, promotion of creativity and innovation among students and teamwork and communication. The chapter ends by highlighting the necessity of systematic and well-planned transition of ICT through DBL, and the considerable role of continuous teacher support and professional development for effective ICT integration. In a nutshell, this book chapter presents how DS facilitates the process of the integration of ICT into education curricula using DBL, outlining advantages and challenges of this approach for educators and students.

Keywords CT Integration · Design-based learning · Digital storytelling · Considerations for CT integration

U. F. Ermiş (✉)
Department of Educational Sciences, Amasya University, Amasya, Turkey
e-mail: uf.ermis@amasya.edu.tr

H. C. Balıkçı
Siverek Vocational School, Harran University, Urfa, Turkey
e-mail: hcelal@harran.edu.tr

M. Saritepeci and H. Yildiz Durak (eds.), *Integrating Computational Thinking Through Design-Based Learning*, https://doi.org/10.1007/978-981-96-0853-9_8

Introduction

Since the mid-twenty-first century, computational thinking (CT) has emerged as a basic skill, paralleling traditional literacy and numeracy skills (Andrian & Hikmawan, 2021). Its integration is crucial for fostering abilities in teamwork, analytical reasoning, creativity, and problem-solving (Su & Yang, 2023). The broad applicability of CT, applicable beyond computer science to fields like the natural sciences, humanities, and arts, enhances its significance (Yadav et al., 2014). Through CT's core processes—breaking down complex issues, recognizing patterns, omitting non-essential information, and formulating systematic solutions—individuals learn to tackle intricate challenges effectively (Wing, 2006). Moreover, CT encourages the exploration of digital innovation, enabling users to advance in various pursuits (Grover & Pea, 2013). One of the important contributions of CT is that it equips students with skills for the future and helps them prepare for a wide variety of careers in the digital age, including those that do not yet exist (Barr & Stephenson, 2011).

A method for integration CT into learning practices is by using Design-Based Learning (DBL). Built on the constructivist theory of learning, explained by Papert (1980), DBL endorses the construction of knowledge through experiences and interactions with the world. DBL is successful in developing students' information processing skills by engaging students in problem-solving and design-based activities. Thanks to DBL activities, the students will be able to use CT concepts in real life situations. By doing so, DBL activities enable students to understand and apply CT concepts. Further, when it comes to CT integration, DBL is an open-ended model that let students test and refine their own designs.

There is evidence in the literature that DBL has been practical abd useful in improving CT skills. Evidence comes from Dochshanov and Tramonti (2020) who concluded that students engaging in DBL activities developed their problem CT skills., Sengupta et al. (2013), similarly found that in interactive DBL settings, students obtained a profound understanding of CT principles by engaging in designing and problem solving activities. Furthermore, application of DBL approach in programming education was found by Sarıtepeci (2020a) to have a positive correlation with the growing of CT competencies.

DS embraces a form of digital media for creating narratives or informational content, which is integral to DBL methodologies that target real-life problems using iterative design processes. This synthesis encourages active involvement in learning making students creators and creating the depth of the understanding of the material (Robin, 2008). Students are engaged in critical thinking and problem solving, which are the central components of design-based learning, in order to communicate the messages, they want to express in their digital stories. Moreover, digital stories making improves students' abilities to operate digital tools and technologies. (Robin, 2008).

Embedding CT skills into the curriculum using design-based learning is most practically achieved with DS activities. The Partnership for 21st Century Learning

(P21, 2009) categorizes essential 21st-century skills into three broad domains: The given section of the book contains the data about the ways that CT principles can be combined with DS techniques in the development of critical sub-skills outlined by P21—"Creativity and Innovation," "Communication and Collaboration," "Information, Media, and Technology Skills," "Information Literacy," "Media Literacy," and "Productivity and Accountability." In this context, in this part of the book, we have provided information on how CT concepts can be integrated into DS processes.

Designing a Digital Storytelling Activity

Storytelling has been one of the most effective ways of sharing human experiences, emotions, and thoughts for centuries, not only entertaining but also inspiring (Linaza et al., 2004). Nowadays when digital technology is developing, storytelling gets a new digital face, completely different from the traditional ones (Konaş & Durak, 2023; Smeda et al., 2014). It evolved from blending the traditional storytelling methods with digital tools, therefore making it a powerful platform for promoting ideas and driving the audience to action through a creative utilization of digital media (Shelton et al., 2017). The applications of this strong means of communication are not identical in different fields and these applications are backed up by numerous studies (Barber, 2016; Gubrium, 2009; Liu et al., 2018; Robin, 2008).

Barber (2016) relied on DS in art, Gubrium (2009) used the DS in health, Liu et al. (2018) embraced DS in language education, while Robin (2008) evaluated the DS effectiveness through the TPACK framework in classroom settings. This communication device appears to be a creative innovation in contemporary education (Sabari & Hashim, 2023). It has been noted to be good at combining instructional content with learning activities and in creating conducive learning environments (Smeda et al., 2014). Researches emphasize positive effect of DS on teaching foreign languages and improvement of the linguistic abilities of the students (Yuliani & Hartanto, 2021).

It is also stated that it is effective in increasing students' motivation, satisfaction and participation (Adara, 2022; Hava, 2021; Liu et al., 2018; Sarıtepeci, 2020b; Yildiz Durak, 2018). Beyond language proficiency, DS is also seen as a support mechanism for twenty-first-century skills (Liu et al., 2018; Niemi & Multisilta, 2015), contributing to the advancement of digital literacy (Churchill, 2020), creativity (Lambert, 2013), critical thinking (Stork, 2020), and CT (Lee et al., 2014).

In educational settings, DS appears in linear and non-linear forms (Liu et al., 2010; Özüdoğru & Çakır, 2021b; Spaniol et al., 2006). Linear digital stories unfold along a fixed trajectory, where the narrative's course is set in advance, offering viewers limited interaction—primarily pausing or progressing the story, whereas non-linear DS empowers users with greater narrative control, presenting multiple pathways and endings (Özüdoğru & Çakır, 2021a). This storytelling is often used in gamified stories and gives the user the chance to influence the flow of the story by offering different options (Letonsaari & Selin, 2018). Game stories often involve non-linear processes where the player's choices can lead to different outcomes (Mitchell & McGee, 2011). Likewise, hypertext documents and interactive tales facilitate a non-linear consumption, allowing readers the freedom to navigate through the text non-sequentially (Agosti et al., 1997; Pope, 2020). This aspect aligns "hypertext documents" and "interactive stories" closely with "game stories," showcasing the interactive engagement learners have in the storytelling process.

Integrating CT Concepts into Digital Storytelling

CT is identified as an essential skill that enhances every child's analytical capabilities, providing them with the critical thinking skills applicable across various fields (Voogt et al., 2015). DS projects offer a platform for students to engage in online communication with peers, develop critical and CT, enhance logical reasoning, and gain a deeper insight into the interconnections between knowledge and theoretical frameworks (Lin & Chang, 2021). Furthermore, DS can be used as a tool to apply coding concepts in the real world by improving students' ability to gather information, solve problems and work collaboratively (Tran, 2019).

The International Society for Technology in Education (ISTE) (2024) identifies four core concepts of CT: algorithm generation, pattern recognition, abstraction, and problem decomposition. Notably, there is a parallel between these CT concepts and the stages involved in DS.

Algorithm Creation: The scenario creation phase of DS is similar to algorithm creation. Just as an algorithm is developed through a series of step-by-step instructions to address a problem (Yadav et al., 2016), in DS, the scriptwriter outlines the sequence of events to unfold the narrative in a methodical manner.

Pattern Recognition: Pattern recognition is defined as the automatic recognition of patterns and order in information (Youjun & Xiaomei, 2022). In other words, it can be expressed as being able to distinguish the properties of an object and how it will affect other objects. Similarly, in DS, recognizing patterns in character behaviors, plot development, and other narrative elements is crucial. This ability to identify such patterns aids in forecasting the storyline's direction and grasping the underlying message intended by the narrative.

Abstraction: In the framework of CT, abstraction is about deciding what the person will take into account and what he or she will omit for the issue-solving (Qian & Choi, 2023). This is similar to the process of topics segmentation into subtopics and the presentation of main points in DS which is parallel to the phase of abstraction in CT (Kordaki & Kakavas, 2017).

Problem Decomposition: Problem decomposition can be explained as the process of splitting of complex problems into small parts and working on each part systematically (Shute et al., 2017). This is a DS context, and it is apparent when elements of a story are listed, where a story is divided into scenes, and, for each scene, pieces are identified, such as sound, visuals, and animations, and a task is given for the creation of these elements, as an application of problem decomposition.

The CT impact on the process of teaching foreign languages through DS activities was investigated by Parsazadeh et al. (2021) in their study. The findings of the study reveal that the integration of CT strategies in DS results to a major improvement of students' problem solving, language learning outcomes, and motivation.

It is seen that CT concepts are similar to DS stages. It is thought that integrating these concepts into the stages of DS may enable the DS process to be carried out more efficiently and the messages desired to be given through DS to be given more effectively.

Benefits of Integrating CT into Digital Storytelling

DS in the educational setting notably offers students the chance to craft narratives either individually or as part of a group within a workshop environment (Robin, 2008).

When students create stories independently, they strengthen their creative skills and digital literacy (Kearney, 2013). The use of digital tools including texts, audios, and visuals for their narratives enables them to demonstrate their creativity in parallel with their technical expertise (Dreon et al., 2011). During the storytelling process, the storytelling skills are developed, and hence better communication occurs (Razmi et al., 2014). The development of their stories improves their self-esteem (Sarıtepeci, 2020b), while group-based writing workshops develop their teamwork and collaboration skills (Schmoelz, 2018).

By combining DS with CT, students have an opportunity to participate in data analysis, craft meaningful decisions using patterns and algorithms, and use technology as a tool for practical expression of their thoughts and opinions (Kordaki & Kakavas, 2017). Students are closely engaging in generating and sharing their narratives through the digital resources in the DS process (Lutkenhaus et al., 2020). Such participation promotes the practice of CT competences as they are involved in detailed story elements planning, organizing, and sequencing.

Activities created to improve CT skills and DS activities affect different learner skills separately. In this chapter, we focused on the effects of the integration of CT into DS processes, categorizing the outcomes under three key areas: "Developing Problem-Solving Skills" "Promoting Creativity and Innovation" and "Providing Opportunities for Collaboration and Communication" (Fig. 8.1).

Developing Problem-Solving Skills

There has been strong support about the positive impact of DS on students' problem-solving abilities (Hung et al., 2012; Parsazadeh et al., 2021; Robin, 2016; Tengler et al., 2021). In one study by Vinayakumar et al. (2018), DS was embedded in the Scratch platform for fostering 21st-century skills of the students. This environment

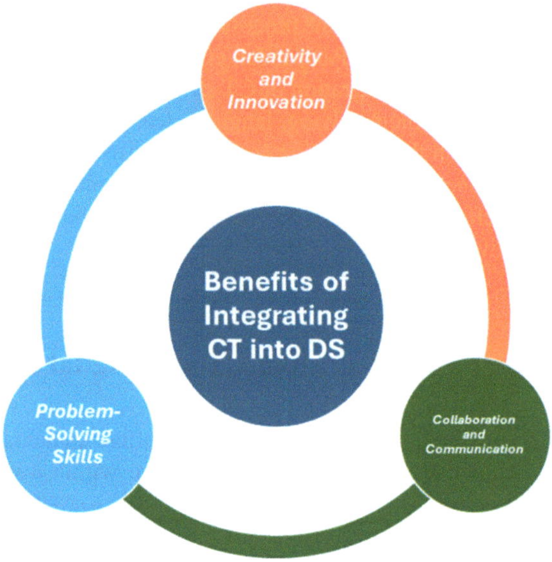

Fig. 8.1 Benefits of integrating CT into digital storytelling

is also noted for being interactive and creative, allowing kids to express themselves without interference. It allows students to bring the ideas to life through creating hard-copy projects, like games, animations, and stories, and to publish them to the wider Scratch online community. The findings of this research point to considerable improvements in students' CT, problem-solving, and reasoning abilities after their participation in this procedure.

CT are practice used to deal with complex problems, including skills of braking the problems down into smaller steps of a solution, determining relevant sources for processing the problem, algorithms designing and implementation, and results interpretation (Wing, 2008). Digital stories are created using CT skills in the following ways: the learner abstracts the story elements, automates the story's progression, and evaluates the story's efficient presentation (Lee et al., 2014).

Integration of CT and DS helps students to consider complex problems as smaller pieces which can be easily solved. Then, they can be solved in a systematic manner (Kordaki & Kakavas, 2017). However, CT e equips students with creativity in identifying and framing problems (Yang et al., 2023), while DS encourage better understanding of problems by constructing stories with defined problems (Yalman, 2023). Therefore, one of the most notable benefits of the integration of CT into DS is the boost of the problem-solving skills of the learners (Parsazadeh et al., 2021).

Promoting Creativity and Innovation

Integrating CT with DS additionally supports innovation and creative thinking (Robin, 2016; Tengler et al., 2021). DS offers learners with several chances to detect, to diagnose and to create innovative ways of solving problems as well as to narrate their own stories (Churchill, 2020; Stork, 2020). Further, DS activities creates a learning environment in which students are allowed to display creativity and diverse thinking (Tengler et al., 2021). Combining CT with DS let learners develop unique stotytelling practices and experiment with different methods and digital devices for the story production (Daily & Eugene, 2013). For instance, it may enable students to use coding languages or multimedia to incorporate interactive features in storytelling and to generate animated versions of stories.

Focusing on English language learners, Yeni et al. (2022) concluded that combining CT with DS developed the students' ability to speaking skills along with their CT skills. The integration approach also helped learners strengthen their digital skills and creativity. DS exercises improve pupils' creativity and writing skills (Sarica & Usluel, 2016). Özüdoğru and Çakır (2020) also found that DS exercises improve creativity and problem-solving abilities among pre-service teachers when used in literacy instruction. Overall, our results suggest that incorporating CT into DS promotes creativity and innovation by enabling students to build their tales using a variety of digital tools and methodologies.

Providing Opportunities for Collaboration and Communication

The activities of CT require a clear interaction in the actions associated with problem-solving. For learners to effectively communicate these stages, they should painstakingly recognize and comprehend the subtleties of the problem, use language precisely, interact constructively with all stakeholders and collaborate towards a solution. Meeting these demands offers learners with the possibility to develop their communication and collaboration skills.

Students are motivated to interact online with peers on the accounts of DS creating a collaborative environment for idea generation (Amaechi et al., 2022). In addition, researches show that DS is used for fostering creative writing and communication skills, which are vital for 21st-century learning (Lin & Chang, 2021).

Evidence has indicated that actions in CT and DS allow learners to enhance their communication and collaboration skills. CT concepts should be a part and parcel of DS projects in order to make these skills improved in a broader sense. Hence, it is rather logical to argue that one of the main advantages of embedding principles of CT into DS processes is that it improves students' collaborative and communicative abilities.

Considerations for CT Integration Through DBL

In line with all integration endeavors, blending CT with DS, especially within a design-based learning framework, requires a methodical approach. To navigate this integration smoothly, certain considerations have been outlined under three key categories.

Thoughtful and Intentional Integration

However, an important consideration of the implementation of the concepts of CT in the design processes in DSDS is thoughtful integration of such an approach. However, this high level of integration requires an extensive knowledge of both DS processes and CT concepts and a judicious use of these two methods together. Furthermore, thoughtful and purposeful integration means choosing design- based learning initiatives that relate to the CT concepts, and, thus, creating an environment where students can engage in these activities (Li, 2014).

CT should be realized as a set of problem-solving methods that are important to computer science but that can be used in many areas (Voogt et al., 2015). Killen et al. (2023) argue the necessity to refocus on better integration of CT and suggest the design of a practitioner-friendly ICT framework that would facilitate its integration into lesson planning. This strategy requires that relevant areas of the curriculum should be mapped where design-based learning and CT approaches can be blended, curriculum envisions, and an ICT framework developed to guide the practice of ICT applications.

Ongoing Professional Development and Support for Educators

Integrating CT into design-based learning activities necessitates a significant commitment to educators' continuing professional development and support. The reason for this is that instructors must have the necessary knowledge and skills in CT, design-based learning, and DS in order to effectively integrate CT principles into their teaching. Three critical components underlie teachers' professional development and long-term support: first, by enhancing teacher training procedures; second, by providing in-service trainings for teachers; and third, by boosting support from educational stakeholders. Yadav et al. (2016) have proposed including ICT-related courses into teacher education programs to increase teachers' CT skills. Professional development efforts are potential avenues for teachers to gain knowledge and skills in integrating Information and Communications Technology (ICT) across disciplinary boundaries (Kelly et al., 2019). Still, it is important for teacher educators to interact with CT experts, have computing resources, and get professional development support for successful integration (Yadav et al., 2016).

Addressing Equity and Access Issues

The issue of equality and accessibility in the context of CT concepts integration into design-based learning DS activities is crucial. Technology and internet connectivity are essential for students who are engaged in DS and CT activities. Although technology is understood to support a more accessible education, access to technology is uneven which further contributes to educational inequalities (UNESCO, 2023). It is during the process of integration that such inequalities in access to digital devices and stable internet connection become more pronounced leading to already existing educational disparities (Warschauer & Matuchniak, 2010).Schools and teachers should provide learning environments where students have equal access to the tools and materials they will need for DS during and outside school hours.

An additional aspect of equity in education encompasses the need for educational practices to be culturally sensitive and inclusive (Ladson-Billings, 1995). Within this framework, to promote equity, educators ought to allow students the autonomy to select story characters and themes of their preference, considering cultural diversity and personal experiences in the DS creation process.

Another crucial element in fostering equity pertains to the assessment of activities. The practices employed in evaluating activities must be conceived with fairness and inclusivity at their core.

It is suggested that educators should use evaluation rubrics that give separate marks for creativity, problem-solving skills, and technical competence when evaluating students' works.

As a result, the principles of CT play a big part in the integration of DS. The acts of addressing equity and accessibility within this area include technological access, cultural sensitivity, and inclusive assessment methodologies.

Activity Plan for Design-Based Learning: Developing Computational Thinking Skills through Digital Storytelling

This seven-step learning activity plan aligns with the CT-integrated DBL framework (Chap. 4) to guide learners through digital storytelling while developing computational thinking skills. The plan encompasses the following steps: (i) preparation; (ii) determining CT and storytelling concepts; (iii) identifying the design problem through storytelling; (iv) enhancing the storytelling idea; (v) storytelling implementation; (vi) storytelling optimization; and (vii) presentation and assessment.

(i) **Preparation**

- **Tools and Resources Pre-Test**:

- Conduct a pre-test to ensure all digital storytelling tools and platforms are functional and suitable for the activity.
- Confirm usability and accessibility for all participants.

- **Tool Introduction and Training**:
  - Organize training sessions to educate learners on using digital storytelling tools and platforms effectively.
  - Develop and distribute user guides and video tutorials for additional support.
- **Supporting Documentation**:
  - Provide learners with access to online resources, including manuals and instructional videos covering the basics of digital storytelling tools.

(ii) **Determining CT and Storytelling Concepts**

- **Concept Integration**:
  - Identify and integrate key computational thinking (CT) concepts with essential elements of digital storytelling, such as narrative structures, character development, and thematic elements.
- **Instructional Content Presentation**:
  - Present scenarios or events related to digital storytelling to help learners identify relevant design problems within the storytelling context.

(iii) **Using Storytelling to Identify the Design Problem**

- **Reflective Ideation**:
  - Encourage learners to reflect on the storytelling context and propose potential design issues or challenges.
- **Feedback and Refinement**:
  - Provide feedback on the proposed design problems, assisting learners in identifying constraints and potential solutions.
- **Limitations and Possibilities**:
  - Help learners recognize the limitations of their design problems and explore possible solutions within the storytelling framework.

(iv) **Enhancing the Storytelling Idea**

- **Research and Brainstorming**:
  - Allow learners to explore context and narrative possibilities using available resources and tools.
  - Facilitate brainstorming sessions to generate creative ideas.
- **Idea Refinement**:

  - Support learners in refining their storytelling ideas through iterative feedback and discussion.
  - Encourage the use of digital tools to enhance and develop their concepts.

- **Sharing and Feedback**:
  - Have learners present their refined storytelling ideas and receive constructive feedback from peers and instructors.

(v) **Storytelling Implementation**

- **Visualization and Prototyping**:
  - Assist learners in creating prototypes or visualizations of their storytelling projects, such as storyboards, script drafts, or digital story prototypes.
- **Tool Utilization**:
  - Guide learners in implementing their storytelling ideas using digital tools and platforms, ensuring effective application of computational thinking principles.

(vi) **Story Optimization**

- **Feedback Integration**:
  - Encourage learners to seek and incorporate feedback from peers, instructors, or external reviewers into their digital storytelling projects.
- **Testing and Evaluation**:
  - Have learners test their storytelling projects to ensure they effectively address the identified design problems and meet objectives.

(vii) **Presentation and Assessment**

- **Project Presentation**:
  - Organize a session for students to present their digital storytelling projects, demonstrating how their storytelling addresses the design problems.
- **Peer Review**:
  - Facilitate peer review sessions where learners assess each other's projects and provide constructive feedback.
- **Reflection and Improvement**:
  - Collect feedback on the learning experience to refine and enhance the activity plan for future iterations.

Conclusion

The importance of CT as a crucial ability in the twenty-first century has been recognized (Andrian & Hikmawan, 2021). This skill has a significant influence on individual abilities in collaboration, problem-solving, and critical thinking across several areas that include reasoning (Hillon-Salgado et al., 2022; Su & Yang, 2023). This study offers insights on the incorporation of CT into design-based learning models, which is a crucial aspect of the process. Learning in DBL is based on constructivist theory, which recognizes learning as a process related to real life experiences (Papert, 1980). It is a technique that allows the integration of CT concepts into day-to-day situations through problem-solving, creative thinking, and design tasks. In addition, there is information that the effectiveness of DBL is in increasing the CT abilities to students (Dochshanov & Tramonti, 2020; Sarıtepeci, 2020a; Sengupta et al., 2013).

DS is a dynamic platform for creating digital narratives that supports an active learning environment that develops the problem-solving and collaborative skills of the students. It serves as a successful tool in design-based learning for execution, as it provides learners with a collaborative platform where they can produce visual narratives. This integrates perfectly with the tenets of design-based learning that provide for the solution of real-life issues and the use of the learned knowledge in real situations (Lin & Chang, 2021).

The stages of DS have striking similarities with the main concepts of CT namely, algorithm creation, pattern recognition, abstraction, and problem decomposition (Kordaki & Kakavas, 2017). In the algorithm creation perspective, just as procedures are formulated step by step to solve a problem, DS involves splitting the narrative into scenes and building them incrementally. Besides, CT as the pattern recognition involves understanding the features of objects and their relations, is very similar to the process in DS when recognizing the patterns and similarities in characters and settings is very essential for conveying the message of the story. Segmenting a story in sub-themes and highlighting crucial details in DS practices is related to the abstraction principle of CT. Also, as in problem decomposition, where complex issues are divided into comprehensible parts to solve systematically, the DS requires the teamwork to combine elements of sound, visual, and animation exposing the interdependence of these two domains.

Incorporating CT concepts into DS processes is recognized for offering three significant advantages: improving problem-solving skills, promoting creativity and innovation and creating situations for working as a team and communication.

Like in any integration processes, there are issues to be addressed in the integration of the CT concepts into DS activities.

A requirement for integration is a thorough understanding and extensive expertise in both fields, necessitating a deliberate and conscientious approach. Furthermore, the achievement of this integration is contingent upon the teachers' professional growth in both the aforementioned disciplines, as well as the continuous assistance provided to address any challenges that may arise throughout the integration process. The provision of this assistance is reliant upon the involvement and endorsement of school leaders and policymakers, highlighting the vital role that active educational leadership plays in the achievement of technology integration projects (Önalan & Kurt, 2020). Ensuring equal and equitable access to integration opportunities for students is crucial in averting the exacerbation of existing inequities and access challenges in the integration process.

Summary of Key Points

This section is dedicated to the combination of CTL principles with design-oriented learning techniques, focussing on DS as a prime example.

Key points:

- DS integration into educational systems is essential for the development of problem-solving, creative and CT skills.
- Ensuring technology provision, implementation of effective assessment mechanisms and use of different resources in educational settings are some of the main approaches to the advancement of equity in digital education.
- The constructivist paradigm inherent in design-based learning offers a structural foundation for embedding CT principles within educational endeavors.
- DS is an innovative learning technique that has the ability to interest students and improve their academic performance. It allows the students to demonstrate their innovation, problem-solving skills and technical knowledge.
- Evaluations in design-based learning have to be fair and comprehensive that should index separate grades for creativity, problem-solving and technical skills. In short, embedding CT into the essence of DS is an important factor for nurturing critical, creative and CT capabilities.

Implications for Future Practice and Research

The integration of computational thinking (CT) concepts into design-based learning activities, such as digital storytelling (DS), significantly impacts educational practice and presents new research opportunities. Teachers are encouraged to incorporate DS activities into the curriculum to develop CT skills, including problem decomposition, algorithmic thinking, and abstraction. These activities foster collaboration, shared practices, and reflection, creating dynamic and interactive classrooms.

From a research perspective, there is ample opportunity to study the impact of combining CT with DS across various educational settings and disciplines. Such investigations can reveal effective methods for integrating these concepts and enhancing pedagogical approaches, ultimately improving student learning experiences. The integration of CT and DS underscores the potential of DS as a tool for creating learner-centered environments, promoting active interaction and participation, and increasing student engagement.

References

Adara, R. A. (2022). Digital storytelling for improving ESPs students motivation and autonomy during Covid-19 pandemic: A conceptual paper. *Journal of English Language Studies, 4*(1). https://doi.org/10.55215/jetli.v4i1.4818

Agosti, M., Crestani, F., & Melucci, M. (1997). On the use of information retrieval techniques for the automatic construction of hypertext. *Information Processing & Management, 33*(2), 133–144.

Amaechi, C. V., Amaechi, E. C., Onumonu, U. P., & Kgosiemang, I. M. (2022). Systematic review and annotated bibliography on teaching in higher education academies (HEAs) via group learning to adapt with COVID-19. *Education Sciences, 12*(10), 699.

Andrian, R., & Hikmawan, R. (2021). The importance of computational thinking to train structured thinking in problem solving. *Jurnal Online Informatika, 6*(1), 113–117.

Barber, J. F. (2016). Digital storytelling: New opportunities for humanities scholarship and pedagogy. *Cogent Arts & Humanities, 3*(1), 1181037. https://doi.org/10.1080/23311983.2016.1181037

Barr, V., & Stephenson, C. (2011). Bringing computational thinking to K-12: What is involved and what is the role of the computer science education community? *ACM Inroads, 2*(1), 48–54. https://doi.org/10.1145/1929887.1929905

Churchill, N. (2020). Development of students' digital literacy skills through digital storytelling with mobile devices. *Educational Media International, 57*(3), 271–284. https://doi.org/10.1080/09523987.2020.1833680

Daily, S. B., & Eugene, W. (2013). Preparing the future STEM workforce for diverse environments. *Urban Education, 48*(5), 682–704. https://doi.org/10.1177/0042085913490554

Dochshanov, A., & Tramonti, M. (2020). Computational design thinking and physical computing: Preliminary observations of a pilot study. *Robotics, 9*(3), 71. https://doi.org/10.3390/robotics9030071

Dreon, O., Kerper, R. M., & Landis, J. (2011). Digital storytelling: A tool for teaching and learning in the YouTube generation. *Middle School Journal, 42*(5), 4–10. https://doi.org/10.1080/00940771.2011.11461777

Gubrium, A. (2009). Digital storytelling: An emergent method for health promotion research and practice. *Health Promotion Practice, 10*(2), 186–191. https://doi.org/10.1177/1524839909332600

Grover, S., & Pea, R. (2013). Computational thinking in K-12: A review of the state of the field. *Educational Researcher, 42*(1), 38–43. https://doi.org/10.3102/0013189X12463051

Hava, K. (2021). Exploring the role of digital storytelling in student motivation and satisfaction in EFL education. *Computer Assisted Language Learning, 34*(7), 958–978. https://doi.org/10.1080/09588221.2019.1650071

Hillon-Salgado, D. F., Fierro, W. A. G., & Molina-Mosquera, J. J. (2022). Documentary review about computer programming for the development of computer thinking, logical-mathematical skills, and scientific skills. *Academic Journal of Interdisciplinary Studies, 11*(3), 20–29.

Hung, C. M., Hwang, G. J., & Huang, I. (2012). A project-based digital storytelling approach for improving students' learning motivation, problem-solving competence and learning achievement. *Journal of Educational Technology & Society, 15*(4), 368–379.

ISTE. (2024, January 10). *Computational thinking competencies*. https://iste.org/standards/computational-thinking-competencies

Kearney, M. (2013). A learning design for student-generated digital storytelling. *The Future of Learning Design*, 73–92. https://doi.org/10.1080/17439884.2011.553623

Kelly, N., Wright, N., Dawes, L., Kerr, J., & Robertson, A. (2019). Co-design for curriculum planning: A model for professional development for high school teachers. *Australian Journal of Teacher Education (Online), 44*(7), 84–107.

Killen, H., Coenraad, M., Byrne, V., Cabrera, L., Mills, K., Ketelhut, D. J., & Plane, J. D. (2023). Teacher education to integrate computational thinking into elementary science: A design-based research study. *ACM Transactions on Computing Education, 23*(4), 1–36.

Konaş, Z. Ş, & Durak, H. Y. (2023). Examination of the effects of digital story applications on digital parenting and technology usage attitudes. *Instructional Technology and Lifelong Learning, 4*(1), 15–47.
Kordaki, M., & Kakavas, P. (2017). Digital storytelling as an effective framework for the development of computational thinking skills. In *EDULEARN17 proceedings* (pp. 6325–6335). https://doi.org/10.21125/edulearn.2017.2435
Ladson-Billings, G. (1995). Toward a theory of culturally relevant pedagogy. *American Educational Research Journal, 32*(3), 465–491.
Lambert, J. (2013). *Digital storytelling.* Routledge.
Lee, I., Martin, F., & Apone, K. (2014). Integrating computational thinking across the K–8 curriculum. *ACM Inroads, 5*(4), 64–71. https://doi.org/10.1145/2684721.2684736
Letonsaari, M., & Selin, J. (2018). Social media integration with nonlinear educational storytelling application. In *2018 17th International conference on information technology based higher education and training (ITHET)* (pp. 1–5). https://doi.org/10.1109/ITHET.2018.8424793
Li, Y. (2014). Research into the computational thinking for the teaching of computer science. In *IEEE frontiers in education conference (FIE) proceedings, Madrid, Spain* (pp. 1–7). https://doi.org/10.1109/FIE.2014.7044465
Lin, C. H., & Chang, Y. Y. (2021). A progressive digital narrative teaching method to improve learning motivation as a lifelong learning skill. *Sustainability, 13*(23), 12991.
Linaza, M. T., Eskudero, H., Lamsfus, C., & Marcos, G. (2004, December). An authoring tool for interactive digital storytelling. In *VAST* (pp. 203–211).
Liu, C. C., Liu, K.-P., Chen, G.-D., & Liu, B.-J. (2010). Children's collaborative storytelling with linear and nonlinear approaches. *Procedia—Social and Behavioural Sciences, 2*(2), 4787–4792. https://doi.org/10.1016/j.sbspro.2010.03.771
Liu, K. P., Tai, S. J. D., & Liu, C. C. (2018). Enhancing language learning through creation: The effect of digital storytelling on student learning motivation and performance in a school English course. *Educational Technology, Research and Development, 66*(4), 913–935. https://doi.org/10.1007/s11423-018-9592-z
Lutkenhaus, R. O., Jansz, J., & Bouman, M. P. A. (2020). Toward spreadable entertainment-education: Leveraging social influence in online networks. *Health Promotion International, 35*(5), 1241–1250. https://doi.org/10.1093/heapro/daz104
Mitchell, A., & McGee, K. (2011). Supporting rereadability through narrative play. In *Interactive storytelling: Fourth international conference on interactive digital storytelling, ICIDS 2011, Vancouver, Canada, November 28–1 December, 2011. Proceedings 4* (pp. 67–78). Springer.
Niemi, H., & Multisilta, J. (2015). *Digital storytelling promoting twenty-first century skills and student engagement.* https://doi.org/10.1080/1475939x.2015.1074610
Önalan, O., & Kurt, G. (2020). Exploring Turkish EFL teachers' perceptions of the factors affecting technology integration: A case study. *Journal of Language and Linguistic Studies, 16*(2), 626–646.
Özüdoğru, G., & Çakır, H. (2020). An investigation into the opinions of pre-service teachers towards uses of digital storytelling in literacy education. *Participatory Educational Research, 7*(1), 242–256.
Özüdoğru, G., & Çakır, H. (2021a). Investigation of pre-service teachers' opinions about using non-linear digital storytelling method. *Kastamonu Journal of Education, 29*(2), 452–459. https://doi.org/10.24106/kefdergi.744216
Özüdoğru, G., & Çakır, H. (2021b). Non-linear digital storytelling: Effect on technology utilisation and writing self-efficacy. *Technology in Society, 67*, 101798. https://doi.org/10.1016/j.techsoc.2021.101798
Papert, S. (1980). *Mindstorms: Children, computers, and powerful ideas.* Basic Books.
Parsazadeh, N., Cheng, P.-Y., Wu, T.-T., & Huang, Y.-M. (2021). Integrating computational thinking concept into digital storytelling to improve learners' motivation and performance. *Journal of Educational Computing Research, 59*(3), 470–495. https://doi.org/10.1177/0735633120967315
Partnership for 21st Century Skills (2009). *P21 framework definitions.* ERIC Clearinghouse.

Pope, J. (2020). Further on down the digital road: Narrative design and reading pleasure in five New Media Writing Prize narratives. *Convergence, 26*(1), 35–54.

Qian, Y., & Choi, I. (2023). Tracing the essence: Ways to develop abstraction in computational thinking. *Educational Technology Research and Development, 71*(3), 1055–1078.

Razmi, M., Pourali, S., & Nozad, S. (2014). *Digital storytelling in EFL classroom (oral presentation of the story): A pathway to improve oral production.* https://doi.org/10.1016/j.sbspro.2014.03.576

Robin, B. R. (2008). Digital storytelling: A powerful technology tool for the 21st century classroom. *Theory into Practice, 47*(3), 220–228. https://doi.org/10.1080/00405840802153916

Robin, B. R. (2016). The power of digital storytelling to support teaching and learning. *Digital Education Review, 30*, 17–29.

Sabari, N. A. S., & Hashim, H. (2023). Sustaining education with digital storytelling in the English language teaching and learning: A systematic review. *International Journal of Academic Research in Business and Social Sciences, 13*(4), 214–231.

Sarica, H. Ç., & Usluel, Y. K. (2016). The effect of digital storytelling on visual memory and writing skills. *Computers & Education, 94*, 298–309.

Sarıtepeci, M. (2020a). Developing computational thinking skills of high school students: Design-based learning activities and programming tasks. *The Asia-Pacific Education Researcher, 29*(1), 35–54.

Sarıtepeci, M. (2020b). Students' and parents' opinions on the use of digital storytelling in science education. *Technology, Knowledge and Learning, 26*(1), 193–213. https://doi.org/10.1007/s10758-020-09440-y

Schmoelz, A. (2018). Enabling co-creativity through digital storytelling in education. *Thinking Skills and Creativity, 28*, 1–13. https://doi.org/10.1016/j.tsc.2018.02.002

Sengupta, P., Kinnebrew, J. S., Basu, S., Biswas, G., & Clark, D. (2013). Integrating computational thinking with K-12 science education using agent-based computation: A theoretical framework. *Education and Information Technologies, 18*(2), 351–380.

Shelton, C. C., Archambault, L. M., & Hale, A. E. (2017). *Bringing digital storytelling to the elementary classroom: Video production for preservice teachers.* https://doi.org/10.1080/21532974.2016.1276871

Shute, V. J., Sun, C., & Asbell-Clarke, J. (2017). Demystifying computational thinking. *Educational Research Review, 22*, 142–158.

Smeda, N., Dakich, E., & Sharda, N. (2014). The effectiveness of digital storytelling in the classrooms: A comprehensive study. *Smart Learning Environments, 1*(6). https://doi.org/10.1186/s40561-014-0006-3

Spaniol, M., Klamma, R., Sharda, N., & Jarke, M. (2006). Web-based learning with non-linear multimedia stories. In W. Liu, Q. Li, & R. W. H. Lau (Eds.), *Advances in web based learning—ICWL 2006* (Vol. 4181, pp. 249–263). Springer. https://doi.org/10.1007/11925293_23

Stork, M. G. (2020). Supporting twenty-first century competencies using robots and digital storytelling. *Journal of Formative Design in Learning, 4*(1), 43–50. https://doi.org/10.1007/s41686-019-00039-w

Su, J., & Yang, W. (2023). A systematic review of integrating computational thinking in early Childhood Education. *Computers and Education Open, 100122*, 1–12.

Tengler, K., Kastner-Hauler, O., & Sabitzer, B. (2021). Enhancing computational thinking skills using robots and digital storytelling. In *Proceedings of the 13th international conference on computer supported education* (pp. 157–164). https://doi.org/10.5220/0010477001570164

Tran, Y. (2019). Computational thinking equity in elementary classrooms: What third-grade students know and can do. *Journal of Educational Computing Research, 57*(1), 3–31.

UNESCO. (2023). *Global education monitoring report, 2023: Technology in education: a tool on whose terms?* UNESCO. https://doi.org/10.54676/UZQV8501

Vinayakumar, R., Soman, K. P., & Menon, P. (2018). Digital storytelling using scratch: Engaging children towards digital storytelling. In *2018 9th international conference on computing, communication and networking technologies (ICCCNT)* (pp. 1–6). https://doi.org/10.1109/ICCCNT.2018.8493941

Voogt, J., Fisser, P., Good, J., Mishra, P., & Yadav, A. (2015). Computational thinking in compulsory education: Towards an agenda for research and practice. *Education and Information Technologies, 20*, 715–728.

Warschauer, M., & Matuchniak, T. (2010). New technology and digital worlds: Analysing evidence of equity in access, use, and outcomes. *Review of Research in Education, 34*(1), 179–225.

Wing, J. (2006). Computational thinking. *Communications of the ACM, 49*(3), 33–35. https://doi.org/10.1145/1118178.1118215

Wing, J. (2008). *Computational thinking and thinking about computing*, 1–1. https://doi.org/10.1109/IPDPS.2008.4536091

Yadav, A., Hong, H., & Stephenson, C. (2016). Computational thinking for all: Pedagogical approaches to embedding 21st century problem solving in K-12 classrooms. *TechTrends, 60*, 565–568.

Yadav, A., Mayfield, C., Zhou, N., Hambrusch, S., & Korb, J. T. (2014). Computational thinking in elementary and secondary teacher education. *Teaching and Teacher Education, 41*, 51–65.

Yalman, F. E. (2023). Is digital storytelling functional in socioscientific issues? Pre-service science teachers' experiences. *Journal of Advanced Education Studies, 5*(1), 203–232.

Yang, W., Ng, D. T. K., & Su, J. (2023). The impact of story-inspired programming on preschool children's computational thinking: A multi-group experiment. *Thinking Skills and Creativity, 47*, 101218.

Yeni, S., Nijenhuis-Voogt, J., Hermans, F., & Barendsen, E. (2022). An integration of computational thinking and language arts: The contribution of digital storytelling to students' learning. In *Proceedings of the 17th workshop in primary and secondary computing education* (pp. 1–10). https://doi.org/10.1145/3556787.3556858

Yildiz Durak, H. (2018). Digital story design activities used for teaching programming effect on learning of programming concepts, programming self-efficacy, and participation and analysis of student experiences. *Journal of Computer Assisted Learning, 34*(6), 740–752.

Youjun, T., & Xiaomei, M. (2022). Computational thinking: A mediation tool and higher-order thinking for linking EFL grammar knowledge with competency. *Thinking Skills and Creativity, 46*, 101143.

Yuliani, S., & Hartanto, D. (2021). Designing digital storytelling in English online learning. *International Journal of Educational Management and Innovation, 2*(3), Article 3. https://doi.org/10.12928/ijemi.v2i3.4139

Chapter 9
Computational Thinking Integration by Design-Based Learning Perspective Implementation Example of Math Education

Deniz Kaya and Yıldız Özaydın Aydoğdu

Abstract This section discusses the importance of computational thinking in mathematics education from the perspective of a design-based learning approach. It explains the universal values of computational thinking and examines studies on its integration into mathematics education. In addition, the characteristic features of computational thinking, which aims to solve problems and involves a cognitive process, have been tried to be explained. In the next step, the prominent studies in the literature on computational thinking were examined from the perspective of mathematics. In this context, the research designs, grade levels, content domain, application style and outcome(s) of mathematical studies in which computational thinking is integrated were examined. Afterwards, design-based learning activities focused on computational thinking for mathematics were mentioned. The design-based learning approach for computational thinking in mathematics education is explained, focusing on activity design, real-world context, collaboration, and reflection. The issues to be considered in design-based learning activities focused on computational thinking for mathematics, tutorial tasks and roles, ways to include real-world contexts, and the development and reflections of the understanding of collaboration are explained. Finally, the text discusses best practice examples in design-based mathematics teaching that focus on computational thinking.

Keywords Computational thinking · Design-based learning · Mathematics education · Problem-based learning

---

D. Kaya (✉)
Department of Mathematics Education, Nevşehir Hacı Bektaş Veli University, Nevşehir, Türkiye
e-mail: denizkaya@nevsehir.edu.tr

Y. Özaydın Aydoğdu
Distance Education Application and Research Center, Nevşehir Hacı Bektaş Veli University, Nevşehir, Türkiye
e-mail: yildizaydogdu@nevsehir.edu.tr

M. Saritepeci and H. Yildiz Durak (eds.), *Integrating Computational Thinking Through Design-Based Learning*, https://doi.org/10.1007/978-981-96-0853-9_9

Introduction

Computational thinking (CT), which represents the indicator of basic thinking skills, includes but is not limited to the skills of formulating, analyzing solutions, logical thinking, abstracting, algorithmic thinking, and generalizing, which require mathematical understanding (International Society for Technology in Education [ISTE] & the Computer Science Teachers Association [CSTA], 2011). In particular, being persistent in solving difficult problems, gaining competence in overcoming open-ended problems, feeling confident in complex situations, being tolerant in dealing with uncertainties, and developing the perception of ability are among the important tendencies of CT (ISTE & CSTA, 2011; Looi et al., 2023). For this reason, it is important that mathematical understandings, which are an important threshold like CT, are supported by tendencies (Calder, 2018). CT skills, which require understanding mathematical thinking, are the common language of many disciplines. It develops by feeding not only from the field of mathematics but also from other disciplines. Thanks to these powerful features, it adds value to students' mathematics learning and paves the way for them to think in multiple ways by supporting their development (e.g., Bartolini-Bussi & Baccaglini-Frank, 2015; Calao et al., 2015; Fanchamps et al., 2021; Fofang et al., 2020; Grover & Pea, 2013; Rodríguez-Martínez et al., 2020; Sáez-López et al., 2019; Sung et al., 2017). Mathematical understanding is always needed for the effective use of CT skills. In this approach, which aims to solve problems creatively and efficiently, one step forward is the ability to think symbolically and conceptually. Intuitive reasoning is among the most needed cognitive skills, especially to achieve results. Dimensions that include cognitive processes include reformulation, recursion, decomposition, abstraction, and systematic testing (Wing, 2006).

One of the most important characteristic features of CT thinking is that it requires the ability to make abstractions (Wing, 2006, 2010). Abstraction is a phenomenon constantly progressing in the science of mathematics, and it involves vertically redesigning previously acquired mathematical knowledge to create a new mathematical structure (Cui & Ng, 2021; Hershkowitz et al., 2001). Abstraction is the process that supports the creation of categories and structuring of a field (Hoppe & Werneburg, 2019). In this way, by reducing the complexity of the problems, it is possible to build larger systems based on the base case (Wing, 2006). Defining the problem by purifying it from the required content, generalizing, and creating patterns can be shown as basic abstraction skills (Kert, 2022). To understand the nature of CT more closely, it is necessary to look at the abilities that would be considered mathematical thinking, the abilities that would be considered CT, and the abilities that require both. Approaches that indicate the relationship between mathematical thinking and CT consist of problem-solving, modeling, data analysis and interpretation, and statistics and probability skills (Sneider et al., 2014). These skills include CT and mathematics-related skills while emphasizing the importance of combining CT and mathematics.

Barcelos et al. (2018) stated that CT in mathematics includes higher-level skills such as semiotic representations (e.g., charts, tables, formulas, verbal or algorithmic representation of a solution), identifying patterns (e.g., analogy, generalization, and specialization) and building models (e.g., student should be able to use mathematic or algorithmic language to build models that explain familiar situations, such as profit and loss, kinematics, opinion polls, and many other) (p. 819). In the report prepared by the National Research Council [NRC] (2010), the basic and universal values of CT among fields are hypothesis testing, data management, parallelism, abstraction, and debugging. This definition indicates that CT is the product of a mathematical thinking process. In this context, CT includes the skills of algorithmic thinking, algebraic thinking, problem-solving, probabilistic thinking, data editing, abstract thinking, inference, modeling, and testing the accuracy of the problem. Shute et al. (2017) also remind us of the need for mathematical skills in solving a complex problem. Gadanidis et al. (2017) emphasize a natural connection between CT and mathematics regarding logical structure and the ability to explore and model mathematical relationships. Therefore, CT has an important place in mathematics and is used frequently. It plays an important role, especially in reconstructing a complex mathematical problem-solving process and shaping the mathematical perspective.

An Overview of CT Integration in Math

Weintrop et al. (2016) further expanded CT and developed a taxonomy divided into four main categories; data handling, modeling and simulation, computational problem solving, and systems thinking to make it suitable for the increasingly computational nature of modern science and mathematics. The need for a taxonomy is illustrated by the increasing presence of computation in mathematics and scientific contexts and the lack of a theoretical basis for what form it should take in school lessons. Although CT-related research has found application opportunities at many educational levels, it has received more response in fields related to K-12 education (e.g., Grover, 2017; Jocius et al., 2021; Kafai, 2016; Sırakaya et al., 2020; Ye et al., 2023). More widespread adoption of the use of CT-based mathematics teaching approaches in K-12 contexts (e.g., Ng et al., 2023; Pei et al., 2018; Ye et al., 2023) has also increased trends towards design-based research, and developing the conceptual basis and application of computationally enhanced mathematics education has become the primary goal (Ng & Cui, 2021, p. 848). These studies also include technology-based instructional designs that make mathematics learning more exciting and accessible. CT can be viewed as a learning tool and object, such as problem-solving and thinking skills for positive disposition (Wu & Yang, 2022). Kotsopoulos et al. (2017) proposed CT pedagogical experiences that include computerized teaching activities in mathematical problem solving, including unplugged, tinkering, making, and remixing, developed from social-constructivism and constructivism theories. In addition, geometry and measurement (Miller, 2019; Pei et al., 2018; Sinclair & Patterson, 2018), numbers and modeling (Chan et al.,

2021; Sung & Black, 2021), and algebra (Bråting & Kilhamn, 2021; Sanford, 2018) are other prominent mathematical studies. The literature study results show that most empirical articles focus on aspects that require mathematical language in the form of automation, abstraction, modeling, algorithmic thinking, visualization, parsing, and pattern recognition (Kallia et al., 2021). Information on the content of some studies involving mathematical understanding is shared (Table 9.1).

These studies indicate that CT application further deepens mathematical discipline knowledge, improves CT skills, and significantly increases the interaction between computer science and mathematical concepts (Looi et al., 2023; Pei et al., 2018; Ye et al., 2023). However, although programming is seen to be influential in the development of specific mathematical understanding, it is stated that it cannot be concluded that all areas of mathematics are suitable for learning (Israel & Lash, 2020; Kafai, 2016; Lockwood & De Chenne, 2020). Especially the compatibility between problem-solving and CT and the lack of pedagogical teaching strategies with CT are the difficulties encountered (Ng & Cui, 2021; Refvik & Opsal, 2023; Rodríguez-Martínez et al., 2020). Similarly, it has been stated that students encounter various difficulties in solving mathematical problems in programming contexts due to potential differences between computational and mathematical thinking (Cui et al., 2023). Therefore, there is a lack of consensus in the educational context regarding the relationship and overlap between CT and mathematics.

CT-Driven Design-Based Learning Activities for Mathematics Education

Design-based learning (DBL) activities enable students to actively participate in learning (Driscoll, 2001) and motivate students to learn (Doppelt, 2003). This participation aims to provide students with experience. A student who gains new experiences and actively participates in learning makes sense of the information. Using DBL to make sense of information improves the student's cognitive and metacognitive skills, such as taking personal responsibility, self-esteem, and self-confidence (Waks, 1995). To develop these skills, teaching must first be designed within the framework of outcomes. Following the design, unstructured real-life problems should be created for the outcome. Students must work collaboratively while solving these problems to witness different experiences and ideas they cannot think of themselves. As a result of real-life problems solved through collaboration, students can reflect on their experiences. In this way, how the students' experiences are internalized while solving problems is evaluated.

Table 9.1 Computational thinking studies with mathematical content

Study	Research design	Grade level	Content domain	Application	Outcomes
Rodríguez-Martínez et al. (2020)	Quasi-experimental	Middle school	Numbers	Scratch unplugged	Mathematical ideas and computational thinking
Costa et al. (2017)	Quasi-experimental	Middle school	Probability and statistics notions	Math questions	Developing problem solving skills
Calao et al. (2015)	Quasi-experimental	Middle school	–	Scratch	Understanding of mathematical knowledge
Benton et al. (2018)	Case study	Primary school	Numbers	Scratch	Facilitates engagement and enthusiasm
Bartolini-Bussi and Baccaglini-Frank (2015)	Teaching experiment	Pre-school	Geometry	Robot Program	Facilitates mathematical meaning
Calder (2018)	Case study	Primary school	Geometry	Scratch	Facilitates mathematical thinking
Fanchamps et al. (2021)	Exploratory study	Middle school	Numbers	Lego Program	Learning efficiency and the learning effect
Fofang et al. (2020)	Video-Interviews	Elementary school	Numbers	Sphero. Math activity	Integrated CT into classrooms in effective and equitable ways
Gadanidis et al. (2018)	Case study	Middle school/Teacher	Algebra	Scratch and Google's blocky	Opportunities for mathematical conceptual insight
Israel and Lash (2020)	Document analysis	Elementary school	Geometry numbers algebra	Scratch and code.org unplugged	Design integrated instructional units for elementary classrooms

(continued)

Table 9.1 (continued)

Study	Research design	Grade level	Content domain	Application	Outcomes
Miller (2019)	Experiment methodology	Elementary school	Algebra	Scratch	Development of mathematical thinking in patterns-structures
Refvik and Opsal (2023)	Quasi-experimental	Middle school	Problem test (PISA)	Optional program	Increase in student score
Sáez-López et al. (2019)	Mixed and complementary	Primary school	Geometry numbers	mBot	Understanding of mathematical and computational concept

Activity Design

When designing CT-driven DBL, the instructional designer awaits an activity design process that requires lengthy and detailed work. All factors that will affect the learning process before, during, and after learning should be considered and the activity design should be made according to these factors (Fig. 9.1).

Inactivity design, first, the mathematical learning outcome to be addressed should be determined. These outcomes can be selected from relevant curricula in formal learning environments. If an extra-curricular outcome is to be determined in learning

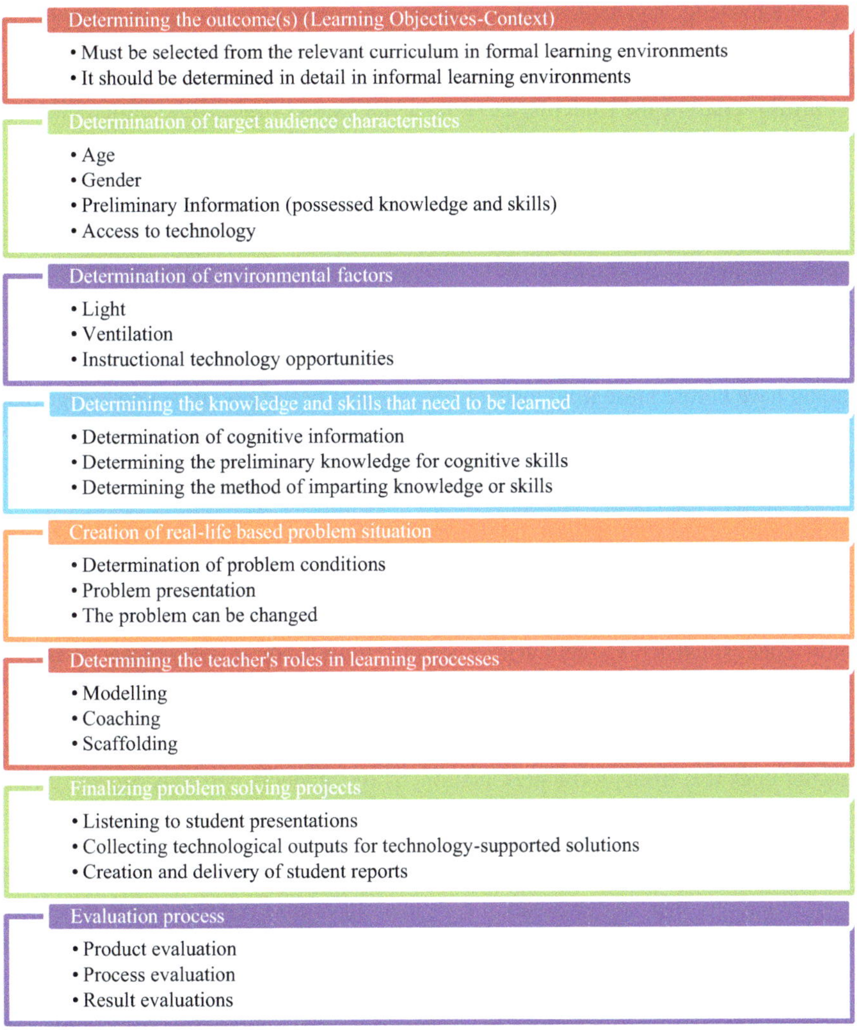

Fig. 9.1 CT-Driven DBL processes to be considered in activity design

environments, the outcome must be written correctly. After the determined outcome, it is necessary to determine which age group this outcome belongs to in the curriculum, which age group the extra-curricular outcomes are created for, and the characteristics of this age group. In addition to age, the fact that the problems are interesting and in the areas of interest for both male and female students will cause the student to be willing to solve this problem. Among the target audience characteristics, the level of prior knowledge is also of great importance. For example, if it is aimed to have students use a computer-aided application in the process of solving a problem prepared for an outcome in a mathematics course, information such as the student's knowledge of this program and their access to this program should be determined in advance, and different designs should be made according to their knowledge or lack of knowledge and access status. Students' access to or use of technology should also be considered in teaching CT skills. Technology-supported software to be used for access opportunities should be selected according to the technological device or infrastructure of the students. Environmental factors (light, ventilation, teaching technology facilities, etc.) are another factor that should be taken into consideration.

After determining the target audience characteristics and environmental factors, students' cognitive knowledge or preliminary knowledge of cognitive skills should have been determined. Designs should be made for the scope and presentation of this information. In this process, the teacher generally explains cognitive information through the presentation, as students cannot learn this information on their own, and it is necessary for them to solve the problem. After determining the target audience characteristics and environmental factors, the cognitive knowledge that the students should have before presenting the real-life problem or the cognitive knowledge that they should have in advance to use the cognitive knowledge is determined. Designs should be made for the scope and presentation of this information. Before presenting cognitive information, any deficiencies in the preliminary knowledge that students should have should be eliminated, or this information should be reminded using methods such as question and answer, demonstration, problem-solving. For example, if the student needs to solve a problem for the determined outcome using the scratch program, students need to know how to use the scratch program as preliminary knowledge. The teaching method necessary to eliminate these deficiencies should be determined. Following the deficiencies in the necessary prior knowledge, the method for teaching the necessary cognitive knowledge in line with the learning goal should be determined. In the process of presenting cognitive information, generally, the teacher does this through the presentation, as the students will play an active role in the next stages, cannot learn this information on their own, or this information is necessary for them to solve the problem.

When creating a real-life-based problem situation, the problem must be created from situations that students can see in their lives and problems that they may encounter in daily life. For example, it is aimed to teach students units such as grams and milliliters. In that case, the teaching problem can be created with problems that they may encounter in daily life, such as "making filter coffee for 3 people" rather than measurements made with beakers or weights in a laboratory environment. In this case, students should measure the grams of coffee per person in grams and the

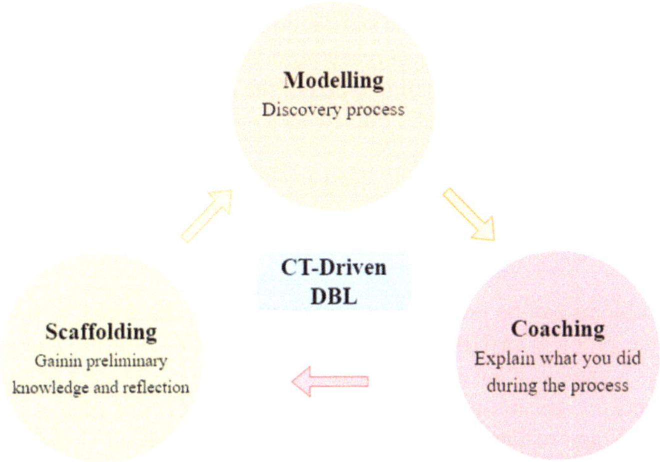

Fig. 9.2 Teacher's duties during student learning activities

amount of water added to this amount per person in milliliters. Rather than learning with beakers or weights, the experience students make during the coffee process, which they will have the opportunity to reinforce with every coffee they make in the future, will be more meaningful. The teacher's roles in design-based activities are modeling, coaching, and scaffolding, as in the constructivist learning approach (Jonassen & Rohrer-Murphy, 1999). The teacher should use these features during the students' learning activities (Fig. 9.2).

The teacher should model the students by solving similar problems while discovering the problem. In this process, the teacher should solve the parts the students have or are thought to have difficulty in without solving the entire problem. Teachers should coach their students by asking questions about their position in problem-solving processes or reminding them of different possibilities through guidance. The teacher should scaffold at the beginning or throughout the process by answering students' questions, presenting the preliminary information that should be obtained before the process, equalizing the knowledge levels by supporting students whose knowledge levels are different, and providing direct explanations. After the problem-solving process, the projects are finalized. At this stage, presenting the processes or products for the solutions created by the students, delivering the outputs of the product obtained through technology-based tools that receive support for the solution of the problem, and preparing the process by the student. It should include an analysis of the problem, an introduction of the product developed to solve the problem, detailed writing of the product outputs, and information about which problems are thought to be solved in the future with this developed product (generalization).

After the problem-solving process, the projects are finalized. One of the most challenging stages of DBL is the evaluation phase. Because, as in other learning methods, it is necessary to evaluate the result, product, and process. In the evaluation phase, the product developed to solve the given mathematical problem is evaluated in terms of effectiveness, efficiency, and satisfaction. For this purpose, teacher observations, individual evaluations, project result reports, or peer evaluations can be used. The result can be made by evaluating whether the desired knowledge or skill level has been achieved within the goal's framework.

Real-World Context

Real-life problems can be prepared as structured, semi-structured, or unstructured. However, the problem in DBL must be semi-structured or unstructured; in other words, it has a variety of answers (such as $2 \times 2 = 4$). For example, in Turkey, 5th-grade students are given a real-life-based problem for the outcome of "relates the area of squares and rectangles with addition and multiplication" which is included in the mathematics curriculum; "I am a painter. I just made a deal to paint the walls of a house. I want to paint the walls. However, I do not know how much paint to buy. I need this calculation every time". When we present the situation "Can you develop a software that calculates the amount of paint I need?", it is seen that the conditions such as the number of walls in the problem, whether the walls contain windows or not, or which coding language or tool will be used are not structured. This situation is open to the student's interpretation. Since these problems are more complex than others a single person cannot overcome them. Therefore, unstructured problems are problem types that can be used in collaborative processes. For a student to solve the problem in DBL, the problem situation must be presented in a semi-structured form. Our semi-structured problem statement should be as follows: "I am a painter. I just made a deal to paint the walls of a house. I want to paint the walls. However, I do not know how much paint I should buy". I need calculations every time. "Can you develop scratch software that tells me the amount of paint needed per wall when I enter the number of walls and the number of windows on the walls?" If we look at the difference between the two problem statements, the conditions of the problem (such as the number of walls, the program to be developed) are not included in the unstructured problem statement.

In contrast, these conditions are included in the semi-structured problem. Determining the conditions is of great importance in DBL. Structured problem statements are not preferred in design-based research because they contain information such as all the details and calculation methods and enable the student to develop the program directly without the need for any research or thinking process. After determining the problem situation, the problem situation can be presented to the students through storytelling, as given in the example above. During the problem presentation, the teacher can demonstrate by pointing to a wall in the classroom. Suppose the context of the problem situation is not included in the classroom. In that case, the teacher can present it to the students using technology-supported visualizations so that the students can visualize the situation.

The changeability of the problem situation means that the learner must be able to change the problem to participate in the process actively. In other words, the problem must achieve the desired result by following different paths. For example, in the paint sample, the student must inform the software of the amount spent per square meter on the wall. The student can make this notification at the end or the beginning of the field calculations. It can change the calculation sequences in the developed algorithms or perform calculations using different methods. Differences in the processes used by students reveal the effect of the process on the result.

Collaboration and Reflection

Collaboration means working with a group or team to gain high-level cognitive skills (Johnson & Johnson, 1990, 2008). Unstructured problem statements require different perspectives and metacognitive skills. Therefore, unstructured problems require collaboration in CT-driven DBL environments. When creating collaborative groups, care should be taken to ensure that the groups are similar in terms of knowledge and skills. To achieve this situation, students' previous experiences should be taken into consideration. Groups can be formed by the teacher, considering the students' previous experiences. However, one of the methods that can be used to minimize intra-group conflicts is for students to form their groups. In design-based research, it is the other essential to determine the time to be given to the groups in case the problem-solving process is carried out with collaborative groups in the activity design. Groups go through stages: Form, Storm, Norm, and Performance (Egolf, 2013). In the forming phase, the group comes together (the formation phase of the group), and in the storm phase, individuals get to know each other's strengths and weaknesses and realize the division of labor. The norm stage is the stage where group rules are determined and discussed. The performance phase is when the necessary work is carried out to achieve the goal (in this case, to solve the real-life problem).

According to Johnson and Johnson (2008), groups should consist of at least two or three students and a maximum of four. However, to reach a consensus in case of group disagreements, it is recommended that these numbers be created from odd numbers of three or more. Odd numbers should be used to avoid disagreements with even numbers. For example, when two people disagree in a group of three people, the third person will be the one who will provide the majority and put an end to the disagreement. However, these numbers should be decided by considering the number of people in the entire group. After the groups are formed, to keep the change in group dynamics under control and to carry out coaching operations by the teacher, the groups should be asked questions and indirect questions asking them to indicate their level of progress in the Form, Storm, Norm, and Performance stages. Depending on the group's situation, the teacher should give the necessary feedback in terms of time and ensure that the groups complete the stages they need to go through. In DBL environments designed collaboratively and including problem-based learning, the reflection phase begins when the problem-solving process is completed in the group's performance phase.

In the reflection phase, students are asked to reflect on both the process and the content they have learned. These reflections include what she/he learned in the process, how he/she thought the process improved her/him, what difficulties she/he experienced in the process, her/his evaluation of the process about her/his friends with whom she/he collaborated, the positive or negative features she/he thought his friends added to her/himself, what she/he thought about the design, quality and real-life operability of the product they created. It should include information such as how she/he thinks he can use what he has learned during the collaboration and problem-solving process in the next stage of his life. In collaboration, not only the student's reflections but also the reflections of his/her friends about the student are of great importance. These reflections enable the person to evaluate the characteristics of her/himself that he cannot see from different perspectives. The problem-solving process in the generalization part of CT must be generalized and transferred to be used in subsequent problems.

The problem-solving process must be evaluated for this transfer to be accurate and used in subsequent processes. Mistakes must be prevented before they are transferred to subsequent processes. Otherwise, it may cause the student to develop incorrect strategies for solving subsequent problems, and as a result, the student may drop out due to inability to solve the problem. To avoid such situations, the student should receive feedback about the problem-solving strategy from the teacher at the end of the process and support during the problem-solving process. Creating reflections is also necessary for the evaluation of learning in collaborative environments. Learning in collaborative environments should be assessed using multiple methods (Johnson & Johnson, 2008). Evaluation of learning should be done formally or informally throughout the entire process (Nelson, 2013). These evaluations should consider the teacher's unstructured conversations or discussions with group members, observation of the group during problem-solving, evaluation of the student's individual process report, and each group's formative reports and usability tests.

A Sample CT-Driven Design-Based Learning Activity

Table 9.2 includes the CT-supported DBL environment design that includes the problem-solving phase that can be carried out individually or collaboratively. This example is not a process that has been tried by the authors before. It is designed as an exemplary CT driven DBL activity in mathematics education.

The aim of the CT-driven DBL environment design given in Table 9.2 is to enable fifth-grade students to develop their CT skills in a real-life problem-based learning environment and to associate the area of squares and rectangles with addition and subtraction operations. As part of the "Information Technologies and Software" course, fifth-grade students study coding with the Scratch tool, creating algorithms, writing operations sequentially, and software processes such as query-loop. Considering this knowledge and skills of the students, these tools and conditions are

Table 9.2 CT-Driven DBL activity that can be designed individually or collaboratively

CT-Driven DBL	Individual	Collaborative
Outcome	Relates the area of square and rectangle with addition and multiplication operations	
Presentation of cognitive information	Teaching how to calculate the area of square and rectangle through presentation (scaffolding)	
Problem situations	Semi Structured	Unstructured
	I am a painter. I just made a deal to paint the walls of a house. I want to paint the walls. However, I don't know how much paint I should buy. I need calculation every time. Can you develop scratch software that tells me the total amount of paint needed when I enter the number of walls and the number of windows on the walls?	I am a painter. I just made a deal to paint the walls of a house. I want to paint the walls. However, I don't know how much paint to buy. I need this calculation every time. Can you develop a software that calculates the amount of paint I need?
Presentation of the problem situation	It will be presented using the display method through the wall in the classroom. After the presentation, the teacher will calculate the area of the wall and the windows, subtract the area of the windows from the area of the wall and tell you how much area will be painted. (Modelling)	It will be presented using the display method through the wall in the classroom. Forming groups of 3 students
Problem solving time	2 weeks	3 weeks
In the problem-solving process	Modeling Coaching Scaffolding At the end of the first week, the teacher answers the questions asked by the students	Modeling Coaching Scaffolding Asking the groups about group work sharing and progress at the end of the 1st week At the end of the 2nd week, students make presentations about their progress

(continued)

Table 9.2 (continued)

Ending the problem	Students present their projects at the end of the process Delivery of individual result report and program output	Making a project completion presentation at the end of the 3rd week Delivery of group result report and program output
Assessment of learning	Product evaluation; Whether the developed program gives the desired result or not Process evaluation; Student reflections Project reports Teacher observation report Result evaluation; Achievement test Student project outcomes	Product evaluation; Whether the developed program gives the desired result or not Process evaluation; Student reflections Project reports Weekly group observation reports Peer reviews Result evaluation; Achievement test and student project outcomes

directly given to them in the case of an individual problem. In the case of a cooperative problem, they are asked to discover how they can solve the problem by using this tool. While solving this problem, students can develop their CT skills through algorithmic thinking and real-life problems. Environmental factors that can benefit students whose prior knowledge levels are determined should also be determined. There are computer laboratories in schools. Since students who do not have personal computer access have access to these laboratories in schools, environmental conditions do not have any restrictive features. To solve this problem, students must know how to calculate the area of a square and rectangle and be able to perform addition and multiplication operations. Fifth-grade students have the knowledge and skills regarding addition and multiplication operations. These students should be cognitively taught how to calculate the area of squares and rectangles by scaffolding by the teacher. For this purpose, the teacher will teach through presentation and make solutions on calculating the area of squares and rectangles through a few examples. After the scaffolding process, the teacher will present the problem situation to individual and collaborative students. While the teacher will model the problem presentation for the students who will make individual solutions, they will not model the problem for the group that will work collaboratively so that they can discover it themselves in the process.

By using Scratch software, the formulating problem stage, which is one of the stages of CT skills, will be supported by students. After the teacher presents the problem situation in the cooperative group, the students will form working groups of three students. The teacher created the group formation process rather than left to the students themselves, as they were too young to be aware of group dynamics. During the presentation of the problem situation, students will also be told about the time given to solve these problems. In addition, the teacher will ask the individual group to develop their algorithm designs (abstraction) and ask their questions whenever they want after a one-week study (logical) and to explain their projects in the class (analyzing) and prepare a report at the end of the two weeks. In the collaborative group, all group members will be required to make a presentation on progress (logical) and algorithm designs (abstraction) at the end of the second week and a final project presentation (analyzing) at the end of the third week. The aim of these presentations is for students to see other projects and see the deficiencies they think they have for their following projects, to close them, or to develop different perspectives. The teacher will receive reports and product printouts when individual or collaborative groups' time ends. At the end of the process, the group that solves the problem individually will be asked to reflect on the process, and the collaborative group will be asked to evaluate their other friends in the group and their reflections on the process. Through their reflections, the students' opinions will be taken about what other real-life situations the product they produced for generalization, which is one of the stages of CT skill, could be a solution. The teacher will evaluate the learning. Multiple evaluation methods, such as all reports collected from students, peer evaluation, product output, reflections, observation reports, and achievement tests, should be included in this process. The teacher will grade students by distributing product, process, and result evaluation points.

Conclusion

CT is an essential supporter of mathematical thinking. There are similarities between mathematical thinking and CT, such as problem-solving, modeling, data analysis and interpretation, statistics, and probability. Mathematics teaching driven by CT using DBL brings different learning advantages or skills depending on education levels. In learning environments, stages must be followed when designing CT-supported design-based mathematical learning activities. The first of these stages is the determination of outcomes (learning objectives-learning context). Then, other factors that affect learning, such as target audience characteristics and environmental factors, need to be determined. Following this determination, the knowledge and skills that need to be learned within the determined context must be determined. Then, real-life based problems are created. The teacher must decide whether the problem situation will be solved individually or collaboratively, considering conditions such as the structure of the problem situation (semi-structured, unstructured), student skills, and the solvability of the problem situation. In addition to designing learning in this process, the teacher has duties such as modeling the students' research processes, coaching by explaining what is done throughout the process, providing the necessary preliminary information, and scaffolding with feedback. After creating a real-life-based problem situation, how to end the problem-solving processes should be decided. After the problem-solving process is completed, the phase in which the effectiveness of the learning is evaluated is started. At this stage, a single evaluation method (such as an achievement test) should be avoided, and multiple evaluations should be made. These evaluations should include student-produced products, reports, reflections, presentations, peer evaluation, and the use of achievement tests, as well as product, process, and result evaluation processes.

Summary of Key Points

This section is dedicated to the importance of learning activities by focusing on CT integrated with the DBL perspective in mathematics education.

Key points:

- Integrating CT, which requires mathematical understanding, into the curriculum is necessary to develop the skills of logical thinking, abstraction, algorithmic thinking, analyzing solutions, and generalization.
- CT provides valuable opportunities for meaningful learning experiences. It plays a critical role in improving the potential of learning and teaching mathematics and reshaping our perspective on the educational process.
- The constructivist paradigm inherent in DBL unlocks the power of CT and offers a structural foundation for incorporating it into educational efforts.
- DBL activities provide learning motivation by helping students gain new experiences and supporting their active participation in the learning process. They also support collaborative solving of real-life problems and the development of reflective thinking.
- Assessment and evaluation of design-based mathematics instruction focused on CT should be fair and comprehensive, with separate marks for problem solving (in the context of processes), logical and analytical thinking skills. In short, placing CT at the center of design-based mathematics teaching is an important factor for the development of CT abilities by nurturing reflective, active and collaborative learning.

Implementations for Future Practice and Research

Integrating CT principles and concepts into design-based mathematics learning activities offers different insights into educational practices. This means it will have far-reaching effects on new areas that researchers can focus on in the future. Therefore, instructors should be encouraged to reflect on design-oriented activities that require CT skills such as abstraction, algorithmic thinking, and analysis by disassembling, combining parts, and automation into their learning environments. Such activities not only encourage reflective abstraction skills and a better understanding of our world, but also foster collaborative efforts. Collaborative learning in mathematics can lead to a deeper understanding of concepts, as students can learn from each other's perspectives and approaches. By reducing the complexity of problems, establishing effective connections with other disciplines, and making learning environments dynamic and interactive, these activities can significantly enhance the learning experience. In this regard, it is essential to research and increase the awareness of instructors and program makers to further their understanding of mathematics learning.

Longitudinal studies provide opportunities to examine the effects of design-oriented mathematics teaching integrated with CT in different educational environments and fields of knowledge. These studies may provide more in-depth answers about the pedagogical methods of incorporating these concepts and the best ways to enhance students' learning experiences. Considering that this learning approach centers on active interaction, it may be preferred in teaching mathematical concepts (e.g., logic, algebra, limit, function), which are known to cause difficulties and misconceptions in learning. Designs can be planned in a spiral structure and tested simultaneously at all educational levels. In this way, the impact of CT on mathematics learning areas can be monitored holistically. In addition, CT designs in mathematics education can be considered in different approaches such as flipped learning, APOS (Action-Process-Object-Schema) theory, and realistic mathematics education.

References

Barcelos, T., Muñoz-Soto, R., Villarroel, R., Merino, E., & Silveira, I. (2018). Mathematics learning through computational thinking activities: A systematic literature review. *Journal of Universal Computer Science, 24*(7), 815–845.

Bartolini-Bussi, M. G., & Baccaglini-Frank, A. (2015). Geometry in early years: Sowing seeds for a mathematical definition of squares and rectangles. *ZDM, 47*(3), 391–405. https://doi.org/10.1007/s11858-014-0636-5

Benton, L., Saunders, P., Kalas, I., Hoyles, C., & Noss, R. (2018). Designing for learning mathematics through programming: A case study of pupils engaging with place value. *International Journal of Child-Computer Interaction, 16*, 68–76. https://doi.org/10.1016/j.ijcci.2017.12.004

Bråting, K., & Kilhamn, C. (2021). Exploring the intersection of algebraic and computational thinking. *Mathematical Thinking and Learning, 23*(2), 170–185. https://doi.org/10.1080/10986065.2020.1779012

Calao, L., Moreno-León, J., Correa, H., & Robles, G. (2015). Developing mathematical thinking with scratch: An experiment with sixth-grade students. In G. Conole, T. Klobučar, C. Rensing, J. Konert, & E. Lavoué (Eds.), *Design for teaching and learning in a networked world* (pp. 17–27). Springer Publishing Company. https://doi.org/10.1007/978-3-319-24258-3_2

Calder, N. (2018). Using Scratch to facilitate mathematical thinking. *Waikato Journal of Education, 23*(2), 43–58. https://doi.org/10.15663/wje.v23i2.654

Chan, S. W., Looi, C. K., Ho, W. K., Huang, W., Seow, P., & Wu, L. (2021). Learning number patterns through computational thinking activities: A Rasch model analysis. *Heliyon, 7*(9), 1–14. https://doi.org/10.1016/j.heliyon.2021.e07922

Costa, E. J. F., Campos, L. M. R. S., & Guerrero, D. D. S. (2017, October). Computational thinking in mathematics education: A joint approach to encourage problem-solving ability. *In 2017 IEEE frontiers in education conference (FIE)* (pp. 1–8). IEEE.

Cui, Z., & Ng, O. (2021). The interplay between mathematical and computational thinking in primary school students' mathematical problem-solving within a programming environment.

Journal of Educational Computing Research, 59(5), 988–1012. https://doi.org/10.1177/0735633120979930

Cui, Z., Ng, O., & Jong, M. S.-Y. (2023). Integration of computational thinking with mathematical problem-based learning: Insights on affordances for learning. *Educational Technology & Society, 26*(2), 131–146. https://doi.org/10.30191/ETS.202304_26(2).0010

Driscoll, M. (2001). *Psychology of learning for assessment* (2nd ed.). Allyn and Bacon.

Doppelt, Y. (2003). Implementation and assessment of project-based learning in a flexible environment. *International Journal of Technology and Design Education, 13*(3), 255–272. https://doi.org/10.1023/A:1026125427344

Egolf, D. B. (2013). *Forming storming norming performing: Successful communication in groups and teams.* IUniverse.

Fanchamps, N., Slangen, L., Hennissen, P., & Specht, M. (2021). The influence of SRA programming on algorithmic thinking and self-efficacy using Lego robotics in two types of instruction. *International Journal of Technology and Design Education, 31*(2), 203–222. https://doi.org/10.1007/s10798-019-09559-9

Fofang, J., Weintrop, D., Walton, M., Elby, A., & Walkoe, J. (2020). Mutually supportive mathematics and computational thinking in a fourth-grade classroom. In M. Gresalfi, & I. Horn (Eds.), *The interdisciplinarity of the learning science (14th international conference of the learning sciences)* (Vol. 3, pp. 1389–1396). International Society of the Learning Sciences.

Gadanidis, G., Clements, E., & Yiu, C. (2018). Group theory, computational thinking, and young mathematicians. *Mathematical Thinking and Learning, 20*(1), 32–53. https://doi.org/10.1080/10986065.2018.1403542

Gadanidis, G., Hughes, J., Minniti, L., & White, B. (2017). Computational thinking, grade 1 students and the binomial theorem. *Digital Experiences in Mathematics Education, 3*(2), 77–96. https://doi.org/10.1007/s40751-016-0019-3

Grover, S. (2017). Assessing algorithmic and computational thinking in K-12: Lessons from a middle school classroom. In P. J. Rich, & C. B. Hodges (Eds.), *Emerging research, practice, and policy on computational thinking* (pp. 269–288). Springer Publishing Company. https://doi.org/10.1007/978-3-319-52691-1_17

Grover, S., & Pea, R. (2013). Computational thinking in K-12: A review of the state of the field. *Educational Researcher, 42*(1), 38–43. https://doi.org/10.3102/0013189X12463051

Hershkowitz, R., Schwarz, B., & Dreyfus, T. (2001). Abstraction in contexts: Epistemic actions. *Journal for Research in Mathematics Education, 32*(2), 195–222. https://doi.org/10.2307/749673

Hoppe, H. U., & Werneburg, S. (2019). Computational thinking-More than a variant of scientific inquiry. In S. C. Kong, H. Abelson, & M. Lai (Eds.), *Computational thinking* (pp. 1–10). Springer Publishing Company.

International Society for Technology in Education & the Computer Science Teachers Association (ISTE & CSTA). (2011). *Operational definition of computational thinking for K–12 education.* https://cdn.iste.org/www-root/Computational_Thinking_Operational_Definition_ISTE.pdf

Israel, M., & Lash, T. (2020). From classroom lessons to exploratory learning progressions: Mathematics + computational thinking. *Interactive Learning Environment, 28*, 362–382. https://doi.org/10.1080/10494820.2019.1674879

Jocius, R., Ian O'byrne, W., Albert, J., Joshi, D., Robinson, R., & Andrews, A. (2021). Infusing computational thinking into STEM teaching. *Technology & Society, 24*(4), 166–179. https://doi.org/10.30191/ETS.202110_24(4).0013

Johnson, D. W., & Johnson, R. T. (1990). Cooperative learning and achievement. In S. Sharan (Ed.), *Cooperative learning: Theory and research* (pp. 23–37). Praeger Publishing.

Johnson, R. T., & Johnson, D. W. (2008). Active learning: Cooperation in the classroom. *The Annual Report of Educational Psychology in Japan, 47*, 29–30. https://doi.org/10.5926/arepj1962.47.0_29

Jonassen, D. H., & Rohrer-Murphy, L. (1999). Activity theory as a framework for designing constructivist learning environments. *Educational Technology Research and Development, 47*(1), 61–79. https://doi.org/10.1007/BF02299477

Kafai, Y. B. (2016). From computational thinking to computational participation in K-12 education: Seeking to reframe computational thinking as computational participation. *Communications of the ACM, 59*(8), 26–27. https://doi.org/10.1145/2955114

Kallia, M., van Borkulo, S. P., Drijvers, P., Barendsen, E., & Tolboom, J. (2021). Characterising computational thinking in mathematics education: A literature-informed Delphi study. *Research in Mathematics Education, 23*(2), 159–187. https://doi.org/10.1080/14794802.2020.1852104

Kert, S. B. (2022). Introduction to computer science education. In Y. Gülbahar (Ed.), *From computational thinking to programming* (pp. 1–20). Pegem Akademi Publishing.

Kotsopoulos, D., Floyd, L., Khan, S., Namukasa, I. K., Somanath, S., Weber, J., & Yiu, C. (2017). A pedagogical framework for computational thinking. *Digital Experiences in Mathematics Education, 3*(2), 154–171. https://doi.org/10.1007/s40751-017-0031-2

Lockwood, E., & De Chenne, A. (2020). Enriching students' combinatorial reasoning through the use of loops and conditional statements in Python. *International Journal of Research in Undergraduate Mathematics Education, 6*(3), 303–346. https://doi.org/10.1007/s40753-019-00108-2

Looi, C. K., Chan, S. W., Wu, L., Huang, W., Kim, M. S., & Sun, D. (2023). Exploring computational thinking in the context of mathematics learning in secondary schools: Dispositions, engagement and learning performance. *International Journal of Science and Mathematics Education* (in press). https://doi.org/10.1007/s10763-023-10419-1

Miller, J. (2019). STEM education in the primary years to support mathematical thinking: Using coding to identify mathematical structures and patterns. *ZDM, 51*(6), 915–927. https://doi.org/10.1007/s11858-019-01096-y

National Research Council (NRC). (2010). *Committee for the workshops on computational thinking: Report of a workshop on the scope and nature of computational thinking*. National Academies Press. https://doi.org/10.17226/12840

Nelson, L. M. (2013). Collaborative problem solving. In *Instructional design theories and models* (pp. 241–267). Routledge Publishing.

Ng, O. L., & Cui, Z. (2021). Examining primary students' mathematical problem-solving in a programming context: Towards computationally enhanced mathematics education. *ZDM, 53*(4), 847–860. https://doi.org/10.1007/s11858-020-01200-7

Ng, O. L., Leung, A., & Ye, H. (2023). Exploring computational thinking as a boundary object between mathematics and computer programming for STEM teaching and learning. *ZDM Mathematics Education* (in press). https://doi.org/10.1007/s11858-023-01509-z

Pei, C., Weintrop, D., & Wilensky, U. (2018). Cultivating computational thinking practices and mathematical habits of mind in Lattice Land. *Mathematical Thinking and Learning, 20*(1), 75–89. https://doi.org/10.1080/10986065.2018.1403543

Refvik, K. A. S., & Opsal, H. (2023). Do optional programming courses affect eighth-grade students' mathematical problem solving? *Computers in the Schools, 40*(3), 244–261. https://doi.org/10.1080/07380569.2023.2175634

Rodríguez-Martínez, J. A., González-Calero, J. A., & Sáez-López, J. M. (2020). Computational thinking and mathematics using Scratch: An experiment with sixth-grade students. *Interactive Learning Environments, 28*(3), 316–327. https://doi.org/10.1080/10494820.2019.1612448

Sáez-López, J., Sevillano-García, M., & Vazquez-Cano, E. (2019). The effect of programming on primary school students' mathematical and scientific understanding: Educational use of mBot. *Educational Technology Research and Development, 67*(6), 1405–1425. https://doi.org/10.1007/s11423-019-09648-5

Sanford, J. (2018). Introducing computational thinking through spreadsheets. In M. Khine (Ed.), *Computational thinking in the STEM disciplines* (pp. 99–124). Springer Publishing Company.

Shute, V., Sun, C., & Asbell-Clarke, J. (2017). Demystifying computational thinking. *Educational Research Review, 22*, 142–158. https://doi.org/10.1016/j.edurev.2017.09.003

Sırakaya, M., Alsancak Sırakaya, D., & Korkmaz, Ö. (2020). The impact of STEM attitude and thinking style on computational thinking determined via structural equation modeling. *Journal of Science Education and Technology, 29*(4), 561–572. https://doi.org/10.1007/s10956-020-09836-6

Sinclair, N., & Patterson, M. (2018). The dynamic geometrisation of computer programming. *Mathematical Thinking and Learning, 20*(1), 54–74. https://doi.org/10.1080/10986065.2018.1403541

Sneider, C., Stephenson, C., Schafer, B., & Flick, L. (2014). Computational thinking in high school science classrooms. *The Science Teacher, 81*(5), 10–15. https://doi.org/10.2505/4/tst14_081_05_53

Sung, W., Ahn, J., & Black, J. B. (2017). Introducing computational thinking to young learners: Practicing computational perspectives through embodiment in mathematics education. *Technology, Knowledge and Learning, 22*, 443–463. https://doi.org/10.1007/s10758-017-9328-x

Sung, W., & Black, J. B. (2021). Factors to consider when designing effective learning: Infusing computational thinking in mathematics to support thinking-doing. *Journal of Research on Technology in Education, 53*, 404–426. https://doi.org/10.1080/15391523.2020.1784066

Waks, S. (1995). *Curriculum design: From an art towards a science:* Tempus Publishing.

Weintrop, D., Beheshti, E., Horn, M., Orton, K., Jona, K., Trouille, L., & Wilensky, U. (2016). Defining computational thinking for mathematics and science classrooms. *Journal of Science Education and Technology, 25*(1), 127–147. https://doi.org/10.1007/s10956-015-9581-5

Wing, J. M. (2006). Computational thinking. *Communications of the ACM, 49*(3), 33–35. https://doi.org/10.1145/1118178.1118215

Wing, J. M. (2010). Computational thinking: What and why? The link-*The Magazine of the Varnegie Mellon University School of Computer Science 49*, 1–6. https://www.cs.cmu.edu/~CompThink/resources/TheLinkWing.pdf

Wu, W. R., & Yang, K. L. (2022). The relationships between computational and mathematical thinking: A review study on tasks. *Cogent Education, 9*(1), 1–19. https://doi.org/10.1080/2331186X.2022.2098929

Ye, H., Liang, B., Ng, O. L., & Chai, C. S. (2023). Integration of computational thinking in K-12 mathematics education: A systematic review on CT-based mathematics instruction and student learning. *International Journal of STEM Education, 10*(1), 1–26. https://doi.org/10.1186/s40594-023-00396-w

Chapter 10
Promoting Computational Thinking Through STEM Education

Kemal İzci

Abstract Computational thinking (CT) is a problem-solving methodology that involves breaking down complex problems into smaller, more manageable parts and draws upon principles used in computer science and programming. Problem-solving is an important skill in scientific reasoning, so it is crucial for Science, Technology, Engineering, and Mathematics (STEM) learning in K-12. While CT in STEM context is an understudied area of research, the main question is how to integrate CT rather than why to integrate it in STEM context. Therefore, this chapter, firstly, discusses how CT and STEM are related and support each other as both concepts involve modeling, reasoning, and problem-solving. Secondly, this chapter provides ways to integrate CT in STEM context to promote students learning. For the second aim, the reviewing related research shows that there are three main approaches for integration of CT in K-12 education: (1) introducing programming activities that are separate from content learning, (2) connecting CT with content learning by utilizing problem-solving scenarios to describe, compare, and test predictions about systems, and (3) providing students with insights into how STEM professionals apply CT. In this chapter, I also discuss the advantages and limitations of each of these approaches in supporting CT in STEM context. Regardless of these approaches, the chapter also explains: (a) instructional strategies including problem-based learning, experiential learning and game-based learning, (b) plugged and unplugged instructional tools including Scratch, Lego, card sorting and Robotics and (c) assessments that can be used in STEM context to promote CT skill. Lastly, as the success of any innovation in education depends on how teachers apply it in classrooms, the chapter discusses the requirements that STEM teachers need to possess to successfully integrate CT in STEM instruction.

Keywords STEM · Computational thinking · Technology integration · Teaching of computational thinking

K. İzci (✉)
Department of Educational Science, Necmettin Erbakan University, Konya, Türkiye
e-mail: kizci@erbakan.edu.tr

M. Saritepeci and H. Yildiz Durak (eds.), *Integrating Computational Thinking Through Design-Based Learning*, https://doi.org/10.1007/978-981-96-0853-9_10

Introduction

The proposition of computational thinking (CT) was put forth by Wing (2006) as a methodical approach to cognition grounded in computational principles. CT is a model of thinking that necessary for everyone since the jobs in twenty-first century requires people to use advanced computing technologies in order to be more productive and effective in their fields (Li et al., 2020; Wing, 2006). CT is a problem-solving methodology that involves breaking down complex problems into smaller, more manageable parts and it draws upon principles used in computer science and programming (Wing, 2006). Though there exists limited consensus regarding the precise characterization of CT, it is widely acknowledged within the scholarly discourse that CT encompasses the fundamental concepts of decomposition, abstraction, pattern recognition, and algorithmic thinking (Wang et al., 2022; Wing, 2008).

The rapid development of scientific knowledge confronts people with new and challenging types of problems, which require use of combination of multiple disciplines to provide effective solutions. The use of computing and digital technologies facilitates the framing and solving problems. CT is therefore viewed as an important skill for new generations to understand and solve challenging problems in a systemic way (Wang et al., 2022; Wing, 2008). Although CT has emerged and used in the field of computer science and programming, researchers highlight that CT is a way of thinking and should be integrated into other fields to be more useful (Lee et al., 2020; Li et al., 2020; Wang et al., 2022). In this context, there has been a recent increase in the number of studies on the integration of CT into other fields including education (Park & Park, 2018; Valentine, 2018).

In the educational context, CT has been advocated as a fundamental skill for students to thrive in the digital era (Wing, 2006). It promotes not only the understanding of computer science concepts but also aids in enhancing problem-solving abilities across various domains (Brennan & Resnick, 2012).

The Science, Technology, Engineering, and Mathematics (STEM) education aims to bring real world problems into classroom to promote students twenty-first century skills and content learning; thus STEM provides a natural context for the use of CT skills (Wang et al., 2022). In this line, as the review article of Yeni et al. (2024) showed, the number of studies on the integration of CT in the STEM fields has grown rapidly. As STEM education aims to bring complex everyday problems into the classroom to promote scientific thinking and content learning, these problems can be better understood, analyzed, and solved systemically through the use of CT. Moreover, STEM education and CT are closely related and mutually supportive, as both concepts share common elements such as modelling, reasoning and problem solving (Lee et al., 2020; Weintrop et al., 2016).

Problem-solving is an important skill in scientific reasoning; thus, STEM education requires teachers to engage students into problem-solving processes to promote scientific reasoning skills. Because of the important role of computing in problems solving, National Research Council of the USA highlights "using mathematics and computational thinking" as one of the core practices in the Next Generation Science

Standards document for K-12 students (National Research Council [NRC], 2012). Especially, NRC (2012) notes the role of CT in modelling and expressing the relations among big data sets. While there is not a clear consensus on the definition of CT, some important organizations including International Society for Technology in Education (ISTE) and Computer Science Teacher Association (CSTA) highlight CT as a transferable skills that should be targeted in K-12 levels since CT helps learners to use digital tools to formulize problems, analyze data, develop effective and competent solutions and transfer the gained experiences to other types of problems (ISTE & CSTA, 2011). In addition to the general perspective for CT, researchers have provided discipline-specific practices of CT. For example, Weintrop et al. (2016) provided four categories of practices for integrating CT into STEM education. According to these categories, CT can be effectively integrated into STEM for the following practices: data processing, modelling and simulation, problem solving, and systemic thinking.

Based on the leadership of some important institutions and organizations such as ISTE, CSTA, and NRC, research in the field of CT, including CT in STEM education, is rapidly increasing (Yeni et al., 2024). In addition, many developed nations such as England, Finland, South Korea, the USA, and Australia have engaged in finding effective ways to integrate CT into their curriculum, as the concept of CT is increasingly recognized as an essential skill that individuals must possess in order to be well-informed contributors to society, to excel in STEM fields, and to demonstrate the potential for imaginative problem solving and innovation across a wide range of academic disciplines (Govender, 2022; Grover & Pea, 2018). The increased number of studies and efforts in CT is important as it helps us identify the real benefits of CT for learners, the struggles, and ways to promote CT in education. Since STEM is known as a way to integrate CT concepts into the school curriculum, it would be useful to discuss the essential benefits of integrating CT in STEM education based on the collected studies in this area.

The Advantages of Integrating CT into STEM Education

CT and STEM are closely related to each other as both concepts focus on similar skills (Wang et al., 2022; Wing, 2008). CT helps learners effectively solve any kind of problem by using computing knowledge and digital technologies. Effective implementations of CT support learners' problem solving, analytical thinking, and modeling skills (Aydeniz, 2018; Yu & Guo, 2018). By using CT skills, students are able to approach problems in a structured way, break down complex problems into smaller and more manageable parts, identify patterns, and develop step-by-step solutions. In STEM education, real world problems are used as a context for teaching in order to support the acquisition of core competencies and 21st-century skills. Given that they focus on similar concepts and skills, it is thought that integrating CT and STEM will be beneficial to both. Research findings confirm this claim and show that integrating CT into STEM positively affects STEM learning outcomes (Bounou et al., 2023; Chen et al., 2023; Hava & Koyunlu Ünlü, 2021; Lewis & Shah, 2012; Richardo et al., 2023).

As STEM and CT are interrelated, the benefits of integrating CT into STEM education have become the research interests of different researchers. It was evident that some researchers focused on the effect of integrating CT into one of the STEM disciplines, such as mathematics (Lv et al., 2023), while others focused on the effect of integrating CT into STEM in general (Bounou et al., 2023; Verawati et al., 2023; Yang et al., 2021). Findings from studies that focus on the integration of CT in STEM for student learning outcomes can be grouped into three main themes, including cognitive, affective and behavioral aspects. In terms of cognitive aspects, empirical research has shown that the integration of CT in STEM education improves students' content knowledge and conceptual understanding of science (Aksit & Wiebe, 2020; Basu et al., 2013, 2017; Hutchins et al., 2020; Peel et al., 2019; Yadav et al., 2018), mathematics (Lv et al., 2023; Rodríguez-Martínez et al., 2020), and computer science (Israel & Lash, 2020; Sun et al., 2024) concepts. For the affective aspect, researchers have shown that integrating CT into STEM education increases students' motivation to engage in learning and improves their learning satisfaction (Hutchins et al., 2020; Liao et al., 2022; Lytle et al., 2019), attracts students to choose advanced STEM courses and STEM-related careers (Burgett et al., 2015; Kafai et al., 2014; Matsumoto & Cao, 2017; Swaid, 2015; Urban, 2015; Weintrop et al., 2016), positively influences students' digital self-efficacy and self-exploration (Liao et al., 2022), improves their sense of ownership (Lytle et al., 2019). For the behavioural aspect, researchers have shown that integrating CT into STEM develops twenty-first century skills in learners, including problem solving (Basu et al., 2017; Yu & Guo, 2018), algorithmic thinking, modelling, visual-spatial, analytical thinking, measurement, collaboration (Basu et al., 2017; Miller, 2019; Taylor & Baek, 2019).

In addition to the cognitive, affective and behavioural benefits of integrating CT into STEM for students, it is also important to mention the benefits for teachers. Another benefit of integrating CT into STEM is that it allows all teachers to integrate CT practices into their teaching for all students, from pre-school to university level, so that learners can learn and practice CT concepts. Computer science courses and programming education are considered natural contexts for the development of CT concepts (Bers et al., 2014; Weintrop & Wilensky, 2013). However, these courses are not offered to young students, including pre-school and primary school students, due to the abstract nature of the course content, and these courses are not compulsory in most secondary school programs. Therefore, the integration of CT into STEM activities has the potential to provide opportunities for all students to improve their understanding and practice of CT concepts (Romainor et al., 2018). In addition, the integration of CT into STEM education promotes the achievement of project-based learning. Project-based learning is the most popular context for STEM education as it requires students to be actively involved in the learning process (Saad & Zainudin, 2022). Most elements of CT, or at least four basic core elements of CT, require learners to perform decomposition, recognise available patterns, abstract from existing knowledge and finally design an algorithm to solve problems and ultimately develop new knowledge and products (Hutchins et al., 2020; Weintrop et al., 2016). Since the aim of project-based learning is to enhance student learning through active participation in the learning process based on the learning-by-doing approach, CT plays a facilitating role in this purpose by helping teachers actively engage students in problem-solving processes.

As mentioned above, it is obvious that the concept of CT integrated STEM has many advantages for both students and teachers. Therefore, it is more important to think about how to integrate CT than why to integrate it in the STEM context. There are many studies that focus on reviewing the relevant literature to illustrate the trends, approaches, methods, participants, data collection tools and types of studies related to CT and CT integration into STEM (Govender, 2022; Hsu et al., 2018; Ozyurt & Ozyurt, 2023; Wang et al., 2022; Zhang et al., 2021). These studies are also important; however, we believe that providing the ways, instructional strategies and tools, assessments and efforts to educate teachers about integrating CT into STEM are more useful for both teachers, professional development experts, and researchers. Therefore, this chapter has discussed above how CT and STEM are related and supportive of each other, as both concepts involve modeling, reasoning, and problem solving, drawing on related literature. This chapter will then focus on how we can effectively integrate CT into the STEM context. To do this, we have reviewed the related studies to provide ways to integrate CT in a STEM context to enhance students' learning. The rest of the chapter is organized to provide: (a) main approaches for integrating CT in STEM context for K-12 education; (b) teaching strategies and teaching tools for integrating CT in STEM context, (c) assessment of CT, and (d) requirements that STEM teachers need to have in order to successfully integrate CT in STEM.

Approaches for Integrating CT in STEM Education

The traditional ways of promoting CT in PK-12 depends mostly on computer and coding related courses, while these courses are not compulsory for all students or are not offered to lower-level students such as pre-school children in most countries. Moreover, adding new courses related to CT to the already crowded school curriculum is not considered feasible. Following the introduction of the CT concept by Wing (2006) as a way of problem solving that involves taking advantage of computational principles and computer technologies to solve complex and abstract problems, CT skills are becoming popular as a twenty-first century skill that all students need to develop in order to be successful citizens and workers, as new types of jobs depend mostly on computing technologies. Most countries have therefore tried to find ways to support their students' CT skills as early as possible.

One of the main ways to promote CT skills is to integrate CT concepts and practices into STEM education, as STEM provides a natural context for CT, since problem solving is essential to both concepts. Therefore, it is important to know the approaches, their strengths and limitations, and their examples in order to effectively promote the integration of CT in STEM education. The review of the relevant literature showed us that there are three main approaches for integrating CT into STEM as; programming activities, problem solving scenarios, and application of CT by STEM professionals. Below, I mention what each of these approaches is, how they try to integrate CT into STEM, and what their strengths and limitations are in integrating CT into STEM.

Programming Activities

Programming is the process of making computers do what you want them to do. In order to make computers do the things you want them to do, it is important to know how computers work. To program computers, programming languages such as Python and Java are used to tell computers what to do. While knowing and using the programming languages is important for computer scientists, it is important for everyone to know how computers work and the logic behind computer programming because of the importance of CT skills for current and future job requirements (Wing, 2006). One way of promoting CT in STEM education is to use programming activities in order to teach the concepts and logic of CT and engage learners to practice the programming activities to promote their learning outcomes in STEM fields. There are two ways, known as plugged and unplugged, of using programming activities to integrate CT into STEM teaching. Plugged programming refers to the use of computers and other electronic devices in order to support learners' CT abilities. In plugged activities, generally text-based programming languages such as Java or block-based programming languages such as Lego are employed to develop learners' CT skills. On the other hand, unplugged programing denotes using other things such as cards, puzzles, games, and crosswords to teach coding and CT without relying on computers and other digital tools (Grover & Pea, 2013; Gulbahar et al., 2018). Below, I provide the details, advantages, and limitations of them.

Plugged Programming Activities for STEM

One of the ways to promote CT in STEM is to use plugged programming activities. When we look at programming education, text-based and block-based programming activities are used by researchers to promote CT in STEM education (Kirçali & Özdener, 2023). Text-based programming tools depend on detailed knowledge of a program and require learners to know and use syntax, parentheses, commas and special symbols to control the program to achieve a specific task (Gomes & Mendes, 2007). Thus, text-based programming requires students to know algorithmic and programming logic in order to use a program for a specific task, such as developing a solution to a particular STEM-centered problem. Therefore, in order to integrate CT into STEM through text-based programming, students should first be trained to use a text-based program such as Java before using the program to solve a

STEM problem. While researchers have shown that text-based programming activities improve students' CT and programming skills (Sun et al., 2023, 2024), it is difficult for younger students to use text-based programming activities in STEM education to improve both CT and STEM content knowledge (Kite et al., 2021). Text-based programming offers flexibility and is a productive way for professionals to achieve a desired goal. However, when the population is K-12 students, it is difficult to use text-based programming languages to promote CT in a STEM context. Therefore, researchers rarely prefer to use text-based programming activities to integrate CT in STEM education for younger students (Kirçali & Özdener, 2023; Wang et al., 2022).

On the other hand, block-based programming simplifies the coding process by providing predefined pieces of code. Block-based programming is like a puzzle and requires learners to combine the pieces of the puzzle to achieve their goals. Thus, students who do not know the syntax of programming languages can use block-based programming activities to improve their CT and content knowledge of a specific domain such as science and mathematics (Ching et al., 2018; Sun et al., 2023). Block-based programming, such as Scratch, Alice, and Lego, is easy for students to use and allows students to design their robots, create their animations, and develop their digital storytelling. In other words, block-based programming activities let students to use their imaginations to tell computers to do whatever they want. Furthermore, the use of block-based programming activities engages students to use their skills, such as computational, critical and creative thinking to develop reasonable solutions to a given problem or to develop games, stories and animations to illustrate a system (Aksit & Wiebe, 2020; Bortz et al., 2020). Additionally, designing games and animations via block-based programming takes students' attention to concentrate and supports their motivation (Wang et al., 2022). Therefore, most researchers employ block-based programming tools in order to integrate CT into STEM education to support different learning outcomes such as skill development and achievement of content standards (Kampylis et al., 2023; Wang et al., 2022; Yeni et al., 2024). While the use of block-based programming tools is easier to integrate CT into STEM education, teacher-related factors such as teachers' interests, knowledge, workload, and level of digital literacy, and school related factors such as lack of funds for activities, lack of integrated CT lesson plans, and lack of collaboration among teachers from different disciplines impact their integration into real classrooms to support instruction (Ouyang et al., 2018; Pietros et al., 2022).

Unplugged Programing Activities for STEM

Unplugged programming activities, an approach to supporting learners' CT skills, have recently become popular. In particular, the trend towards making CT accessible to all learners has drawn the attention of researchers to finding ways to support the CT skills of younger students. Unplugged programming aims to support learners' practical knowledge of core CT concepts without requiring learners to know and use the details of computer programs and digital tools (Grover & Pea, 2013; Gulbahar et al.,

2018; Rodriguez et al., 2017). Researchers use different activities such as drawing, problem-solving scenarios, paper–pencil activities, interaction with physical objects, animation, card sorting, and role-playing activities to support students' knowledge and practice of core CT concepts without relying on computer-related tools (Looi et al., 2018). Unplugged programming is mostly preferred for use in early childhood and primary school, as it is difficult for students at these ages to understand the logic and abstraction behind computer programming (Küçükkara & Aksüt, 2021).

CT is a problem-solving approach. According to Caeli and Yadav (2020), people have been solving problems since before the invention of computing tools and technologies. The important components in problem solving processes are people, problems and the tools that people use to solve these problems. Thus, computing technologies and programming activities are tools for people to solve problems. By using these tools, people are able to analyse problems in detail, understand the complexity of the problems and provide effective solutions. The capabilities and functions of these tools provide people with different perspectives and advantages in solving problems. Therefore, it is important to improve the understanding, logic and practice of CT rather than teaching learners specific computational tools (Caeli & Yadav, 2020; Küçükkara & Aksüt, 2021). Once students have acquired a high level of numeracy, it does not matter whether they or the tools they have programmed solve the problems. As a result, teachers need to use the most appropriate tools to meet the needs of learners in order to support their development of CT skills.

Unplugged programming activities are known as technology-free activities and therefore allow the use of low-cost materials to support the integration of CT in STEM education by providing equal opportunities. Furthermore, researchers suggest that starting with unplugged programming activities and then moving to plugged programming is more useful for learners to develop their understanding and practice of CT concepts (Caeli & Yadav, 2020). There are many studies that use unplugged programming activities to integrate CT in STEM education (Barth-Cohen et al., 2019; Feldhausen et al., 2018; Mensan et al., 2020; Montes-León et al., 2020; Ozdinç et al., 2022; Peel et al., 2019; Sung et al., 2017; Tran, 2019). The results of these studies showed that the use of unplugged CT activities in STEM education positively influenced students' content knowledge in science (Feldhausen et al., 2018; Mensan et al., 2020; Peel et al., 2019) and mathematics (Sung et al., 2017). Furthermore, integrating CT into STEM through the use of unplugged activities improves CT skills (Mensan et al., 2020), mathematical reasoning (Montes-León et al., 2020), algorithmic thinking (Çiftçi & Topcu, 2023), modeling skills (Peel et al., 2019), self-efficacy (Çiftçi & Topçu, 2022), collaboration and creativity (Çiftçi & Topcu, 2023). Researchers say that unplugged programming activities support equal access to CT concepts because they don't require computers or digital devices, which may not be available to students from low-income families (Kirçali & Özdener, 2023). These activities also do not require computers and other digital devices, which some teachers find frustrating to use in their teaching.

Problem-Solving Scenarios

One of the ways to support the integration of CT in STEM education is to provide appropriate problem-solving situations. STEM education uses real-world and complex problems as a context for teaching and requires students to develop their STEM content knowledge and skills by developing solutions to these problems (Peel et al., 2019). CT is also known as an approach to the problem-solving process and helps students to systematically understand, analyse and develop solutions to given problems (Grover & Pea, 2013). Therefore, the use of problem-solving scenarios is a natural approach to integrating CT and STEM. In particular, researchers suggest collaborative problem-solving examples as an effective way to integrate CT and STEM, as collaboration and teamwork are considered essential to both concepts (Hong et al., 2021; Zhang et al., 2021).

Researchers in computer science have used problem-solving to support learners' CT concepts and have demonstrated its positive impact on students' learning and enjoyment (Zhong et al., 2016). However, in order to support the integration of CT into STEM, problem-solving scenarios need to provide a context for both CT and STEM practices. Therefore, when selecting or designing a problem-solving scenario, it is important to consider that the scenario allows students to practice and develop their knowledge and skills in CT and STEM. Hong et al. (2021) provided such a problem-solving scenario in their study, as it required students in groups to develop a robot that can help scientists-collect data in an active volcanic area. The results showed that students practised different CT sub-skills, with algorithmic thinking being the most practised skill. Students also reviewed and used the science behind active volcanoes and mathematical concepts such as decimals and divisions. In another study, Pei et al. (2018) used calculating the area of polygons as a problem for high school students to support their CT and mathematical habits of mind. They found that the activity improved students' CT skills and understanding of geometric concepts. Hutchins et al. (2020) also used a problem-solving scenario called 'Delivering medicine to a tribe in a remote Amazon jungle' to support students' learning of kinematics concepts of position, velocity and acceleration and practice of CT concepts. They found that students' learning of content knowledge and CT was improved and they were more motivated to engage in the learning tasks. On the whole, it can be said that the problem-solving scenarios provide an effective context for the successful integration of CT and STEM education.

Using How STEM Professionals Apply CT

One of the reasons for integrating CT into STEM education is that scientists and STEM-related jobs involve computational practices, so students should learn and develop their CT skills in order to be successful in these types of jobs (Denning & Tedre, 2019). In particular, scientists develop their research agenda to solve and provide reasonable explanations for the complex and original problems in their field of study. In order to develop effective solutions to these problems, scientists need to use CT tools and skills that enable them to provide realistic and effective solutions. Furthermore, scientists in STEM fields mostly focus on real-world problems where CT can be used to develop alternative solutions. Weintrop et al. (2016) suggested that "science and mathematics are meaningful contexts in which we can successfully situate the concepts and practices of computational thinking" (p. 128), as they provide a meaningful context for CT practices in STEM fields.

Using CT in practice makes learners see why CT is important for STEM fields and what role CT plays in solving problems in STEM fields. Improvements in computing and digital tools let researchers to develop new types of virtual and real-life experiments, which make scientists to develop innovative solutions to complex problems (Beheshti et al., 2017). Integration of CT into STEM fields helps students enhance their learning in these fields and engage them in problem-solving processes as scientists to prepare them for future economic industries (Barr & Stephenson, 2011; Yadav et al., 2018). Beheshti et al (2017) conducted research to identify the CT practices of professionals in STEM fields. They interviewed 17 professionals working in STEM fields and found that STEM professionals use CT in data practices, systemic thinking practices, modeling and simulation practices and computational problem solving practices. It was also found that researchers relied on experimentation mostly use CT for data practice while theoretical researchers mostly used CT in computational problem solving practices. This is important when providing instructional activities based on how STEM professionals use CT approach to support the integration of CT in STEM education.

Malyn-Smith et al. (2018) provided some examples of the use of CT by STEM professionals. These included ensemble modeling for weather forecasting, the use of computational chemistry to read chemical sequences for drug design, the processing of genomic data in bioinformatics, crowdsourced data in water management systems used in environmental science, and the use of machine learning by scientists to use larger data sets for prediction. These types of examples can be used by teachers to show students how STEM professionals use CT in their work. Students can then engage in similar types of activities to act like a STEM professional, using CT to solve problems or develop systems.

Instructional Strategies, Tools and Assessments for Integrating CT in STEM Education

Instructional strategy is the way in which teachers achieve a desired learning outcome and describes the materials, tools and processes that teachers need to use to support students in achieving learning goals. It is therefore important because it guides teachers in deciding what to do and how to do it. On the other hand, instructional tools are known as the tools and technologies that are used in the classroom for a specific purpose to support the implementation of the instructional strategies. It can be said that instructional tools are the implementation phase of instructional strategies. To effectively integrate CT in STEM, appropriate and effective teaching strategies or models and tools need to be used to plan and deliver the lesson. Below we provided the instructional strategies and tools used by researchers in promoting CT in STEM context and explained the most suggested strategies and tools by experts.

Instructional Strategies

There are many teaching strategies/models used by teachers to support students' learning. These strategies can be broadly categorised into teacher-centred and student-centred strategies. As STEM adopts a student-centred approach to teaching, it is important to use student-centred teaching strategies to support STEM education. CT also includes the concepts and skills related to computation, so it also requires students to actively engage in practice, which can be supported by student-centred teaching. Thus, it can be seen in the literature that researchers mainly suggest the use of student-centred strategies to support the integration of CT and STEM.

STEM provides a natural context for CT and recently many researchers have conducted experimental, theoretical and reviewing studies to support the integration of CT in STEM. Looking at the review studies (Govender, 2022; Hsu et al., 2018; Ogegbo & Ramnarain, 2022; Ozyurt & Ozyurt, 2023; Yeni et al., 2024; Zhang et al., 2021; Wang et al., 2022), it can be seen that different strategies have been used by researchers. They are mainly student-centred strategies and include: experiential learning, problem-based learning, collaborative learning, game-based learning, direct instruction, modelling-based learning, design-based learning, use-modify-create model, project-based learning, authentic learning, inquiry-based learning, exploratory learning, storytelling, active learning, independent learning, place-based design education, embodied learning and situated learning.

There are some teaching strategies that researchers have demonstrated their positive effects and suggested as more suitable for integrating CT and STEM. One of these is problem-based learning. Problem-based learning requires teachers to construct or bring a problem, which can also be real-world problems, into the classroom, and students engage in solving these problems by practising their knowledge and skills in CT and STEM. For example, Nugent et al. (2019) used problem-based learning to engage middle school students in an activity to create a wearable e-textile product. They found that the problem-based learning activity helped students improve their learning of circuitry, programming, engineering design principles, and self-efficacy. Another suggested strategy is experiential learning, which focuses on the notion of learning by doing. Researchers' use of experiential learning showed that when students engaged in designing their own electrical circuits (Kafai et al., 2014), coding to illustrate a system (Lee & Vincent, 2019), developing simulations, and other hands-on activities related to CT and STEM, students' practices and understanding of CT and STEM were improved (Yeni et al., 2024).In addition, game-based learning has also been usefully used by researchers to support CT integration in STEM. Game design easily attracts students' attention and motivates them to engage in learning. Researchers indicated that when students were involved in designing their own games or playing a well-designed game, their knowledge of STEM content and use of CT skills improved (Hoover et al., 2016).

On the other hand, some researchers besides of using one of the student-centered instructional strategies mentioned above to support the integration of CT and STEM, they prefer to use approaches that use a combination of the mentioned student-centered instructional strategies. One of the approaches mostly preferred by researchers in supporting CT and STEM is DBL. DBL is a pedagogical approach based on project-based learning and problem-based learning, and incorporates student-centred principles (Reiser et al., 2024). DBL is used by researchers to promote CT related concepts and skills (Brennan & Resnick, 2012; Saritepeci, 2020). DBL focuses on real-life problems and requires students to decide on their own objectives in order to develop a solution or product to the problems (Doppelt & Schunn, 2008). Thus, DBL seems to be appropriate for STEM and CT as both concepts emphasise real-life problems as a context to support content learning and important work skills.

It is not surprising to see studies using DBL to integrate CT and STEM, given the similarities between the two concepts. In STEM education, the use of design-based approaches is varied. Engineering design processes and scientific inquiry are the two of the approaches used in STEM education as forms of design-based approaches (Ladachart et al., 2024). In order to see the success of DBL in integrating CT and STEM, we need to consider which of the design-based approaches are used in STEM studies to illustrate the outcomes of studies. According to researchers, engineering design, including forward and reverse engineering designs, is most commonly used by researchers as a design-based approach to STEM (Delen & Sen, 2023). The research findings illustrate that the design-based approach is an effective way to integrate CT concepts into STEM (Killen et al., 2023; Liu et al., 2023) and that both students (Li et al., 2021) and teachers (Killen et al., 2023) benefit from the approach.

Instructional Tools

Instructional tools are used by students and teachers during instruction to achieve the intended learning outcomes. The aim of integrating CT in STEM is to support students' expertise in the skills and concepts that are important for both constructs. As mentioned earlier, in the integration of CT in STEM, plugged and unplugged approaches and related tools have been used by researchers during the instructional processes.

The plugged approach to integrating CT with STEM includes text-based and block-based programming tools. The analysis of the review studies showed that researchers mostly used block-based programming tools rather than text-based programming, such as Logo, Python and HTML/CSS, to integrate CT and STEM. In particular, Scratch (Aksit & Wiebe, 2020; Benton et al., 2017; Rodríguez-Martínez et al., 2020) and App Inventor (Papadakis, 2020) were two of the most used block-based programming tools to integrate CT and STEM. Scratch allows students to construct their own model to illustrate how a system works. Such a model gives students the opportunity to control variables within the system, allowing them to demonstrate the relationships between variables (Aksit & Wiebe, 2020). App Inventor allows students to develop their programmes without any programming knowledge. By using these types of block-based programming tools, students engage in the model-building process and deepen their understanding of both content knowledge and CT (Aksit & Wiebe, 2020; Basu et al., 2017; Grover & Pea, 2013; Weintrop et al., 2016). There are different ways to use the plugged programming approach to support the integration of CT in STEM. In addition to coding platforms such as Scratch, robotic kids, simulation software, data visualization tools and gaming tools can be used to integrate CT and STEM. Robotics kits such as LEGO Mindstorms, Arduino or Raspberry Pi provide hands-on experience for students to apply computational thinking in building and programming robots. These kits often include sensors, motors, and controllers that allow students to design and implement solutions to real-world STEM problems (e.g. Sen et al., 2021). Simulation software allows students to model and simulate complex systems, helping them to understand abstract concepts and experiment with different parameters (Weintrop et al., 2016). Examples include NetLogo for agent-based modelling, SimCity for urban planning simulations and PhET Interactive Simulations for physics and chemistry concepts. Educational games and gamified learning platforms incorporate computational thinking concepts into gameplay, making learning engaging and interactive. Examples include Angry Birds for learning gravity concepts in physics through gaming and Kerbal Space Program for understanding physics and engineering principles (Chernenkii, 2020; Weintrop et al., 2016).

The tools used in unplugged approach mostly include the hands-on materials and tools. In particular, these tools let students engage in hands-on, interactive exercises that teach CT concepts without the need for computers or digital devices. These unplugged instructional tools offer diverse and engaging ways to teach CT skills in STEM education settings, fostering creativity, collaboration, and problem-solving abilities without relying on digital technology (Montes-León et al., 2020). These tools let students learn the logic and skills of CT and STEM content knowledge. They can be used by all students but especially are effective for students at the early and primary school levels since it is difficult for them to understand and use programming tools to complete learning activities (Montes-León et al., 2020; Ozdinç et al., 2022). Depending on personal creativity there are various tools that can be developed and used as unplugged tools. For example, physical objects such as card sets can be used to develop algorithms by organizing a set of cards with numbers or letters into the correct order (Peel et al., 2019). Paper prototyping can be another example of unplugged tool. Designing and creating paper prototypes of interfaces, games, or systems helps students understand the iterative design process and encourages problem-solving skills. They can iterate on their designs, make adjustments, and test them with peers (Ozdinç et al., 2022). Manipulatives such as base-ten blocks, fraction tiles, or geometric shapes can be used to explore mathematical concepts like number systems, operations, and geometry. Students can engage in hands-on activities to understand abstract mathematical ideas (Byrne et al., 2023). Puzzles and games are also used as unplugged tools in literature. Logic puzzles, tangrams, Sudoku, and other brain-teasers provide opportunities for students to practice problem-solving, pattern recognition, and critical thinking skills. These activities can be adapted to various age levels and learning objectives. Storytelling and role-playing activities can also be used as unplugged tools. Encouraging students to create stories or act out scenarios related to STEM concepts helps them develop communication skills, creativity, and problem-solving abilities. For example, they can role-play as engineers solving a design challenge or scientists conducting experiments. Lastly, building physical models or conducting demonstrations of scientific principles, such as using balloons to demonstrate gas laws or creating simple machines with household materials, helps students visualize abstract concepts and understand their real-world applications.

Assessments Methods

At the classroom level, assessment can be used to measure learning outcomes after an instructional process has been completed, and it can also be used during the instructional process to support learning. These two purposes of assessment are known as summative and formative assessment. In addition, formal and informal assessments can be used to meet the needs of summative and formative assessments (Black & Wiliam, 2009). When we look at the literature on CT and STEM, we see that various assessment methods such as multiple choice tests, surveys, questionnaires, interviews, portfolios, performance tests and quizzes have been used (Ogegbo & Ramnarain, 2022; Yeni et al., 2024). However, some of the assessment methods have been used by researchers to determine the impact of their interventions for research purposes, and they vary based on the qualitative and quantitative research approaches. We believe it is more important to mention the assessments that can be used by teachers and students in classrooms to assess and support learning of CT and STEM concepts.

As STEM and CT prioritise learning by doing and value hands-on activities, assessments that can be used in this context are more appropriate. Therefore, performance assessment, collaborative assessment and self-assessment tools are considered appropriate assessment methods for evaluating CT-integrated STEM learning. Performance assessment is effective in assessing students' learning outcomes as they engage in performance tasks to use STEM content knowledge and CT skills in developing a project or constructing solutions to a given problem (Barr & Stephenson, 2011). Self-assessment, peer assessment and collaborative assessment are effective for students to evaluate their own, peers' and groups' performance on a task and allow students to see their own progress and mistakes to correct them to deliver a good product and learn from their mistakes (Mendoza Diaz et al., 2022). In addition, portfolio assessment has been suggested by researchers as a good way to assess the learning process by collecting different artifacts and using relevant rubrics (Ogegbo & Ramnarain, 2022). These portfolios can include a collection of student projects, drawings, presentations, reports, written assignments, student reflections and teacher feedback that can be used for formative and summative purposes. In addition, portfolios can be electronic and allow teachers to view different tasks submitted by students, including simulations, games, animations, videos, etc., in order to assess students' learning as products and processes.

STEM and CT are two broad constructs that combine different skills and concepts. It is therefore difficult to assess such broad constructs using only one assessment method. The use of multiple assessments aligned with learning objectives would be more useful in this context. Some assessments can be used to assess CT concepts, whereas others assess STEM content knowledge, and some of them assess students' progress towards specific learning objectives. In this way, the lack of one assessment method can be filled by another. However, as researchers pointed out, the lack of tools for assessing STEM and CT makes it difficult for teachers to assess the multidimensional nature of CT-integrated STEM instruction (Poulakis & Politis, 2021). In the United States, the NRC (2012) announced new standards for science education that included a three-dimensional nature with an emphasis on scientific and engineering practices, crosscutting concepts, and disciplinary core ideas. Subsequently, researchers have attempted to develop and adopt assessments that can be use by teachers to assess and support the three-dimensional nature of learning in science classrooms (Penuel et al., 2019). Such an effort should be made to develop and adopt assessments for teachers to assess their students' CT integrated STEM learning outcomes. A good start can be to analyze the example of the three-dimensional assessments developed for science and then try to find ways to assess CT and STEM concepts and skills in CT-integrated STEM education.

Preparing STEM Teachers to Effectively Integrate CT into STEM Education

The success of any innovation in education depends on how teachers apply it in the classroom, as teachers are the pioneers in bringing new developments in educational fields into the classroom to support student outcomes. Therefore, for the successful integration of CT in STEM education to enhance student learning, teachers need to know what to do and how to integrate CT and STEM into their teaching practices. As Mishra and Koehler (2006) point out, teachers' knowledge of a technological tool does not guarantee the successful integration of the technological innovation into the classroom; teachers also need to know the content and pedagogy for integration. The amalgam of technology, content and pedagogy is referred to by Mishra and Koehler (2006) as TPACK (Technological Pedagogical Content Knowledge) and is proposed as the knowledge base that teachers should develop in order to take advantage of technology to innovate their teaching and improve student learning. However, having the knowledge base for CT integrated STEM is not sufficient for the transition of CT into STEM teaching practices since STEM teachers beliefs, dispositions, and perceived usefulness of CT integration into STEM influence their willingness and efforts to use CT in their STEM instruction (Kite & Park, 2023; Peters-Burton et al., 2023). Furthermore, contextual and infrastructural factors such as supportive school climate and available computer and digital tools can also influence STEM teachers' integration of CT into their classrooms (Kite & Park, 2023).

According to the TPACK framework, teaching CT integrated with STEM should include knowledge of CT tools and technologies (T), recognition of instructional strategies and tools for teaching CT and STEM content knowledge (P), and knowledge of CT concepts and STEM content (content). Therefore, researchers have attempted to develop and provide training programs for pre-service and in-service teachers to support their knowledge base for successful integration of CT into STEM education (Cabrera et al., 2023; Yang et al., 2021). Furthermore, by using the TPACK framework, researchers also provided new frameworks to describe STEM teachers' knowledge base for CT, such as TPACK for STEM teachers (Chai et al., 2019), TPACK-CT (Jocius et al., 2021), and ICTTE (Integrating CT into Teacher Education Project, Cabrera et al., 2023). Researchers have shown that STEM teachers are not ready to integrate CT into STEM education (Kite & Park, 2023; Sands et al., 2018). Firstly, STEM teachers do not have sufficient knowledge of CT tools and technologies (Aljowaed & Alebaikan, 2018; Sands et al., 2018; Yang et al., 2021). Researchers suggested that taking computer science courses, attending workshops and professional development programs, and co-designing processes can be ways to support teachers' knowledge about CT tools and technologies (Peel et al., 2020; Yadav et al., 2018). Secondly, teachers need to be aware of the ways and strategies they can use to integrate CT into their STEM teaching. The results of studies showed that simply learning about the CT tools does not lead teachers to use them in their instruction to support students' STEM learning (Sabo et al., 2022). Researchers have indicated that designing effective professional development programmes that focus on collaboration, co-design and ongoing support are effective in helping teachers learn how to integrate CT into their STEM lesson plans and teaching practices (Dabholkar et al., 2021; Peel et al., 2020; Peters-Burton et al., 2023; Wu et al., 2020). Third, STEM teachers need to know what CT is and the concepts associated with it. The results of studies showed that without CT-specific training, STEM teachers either do not know anything about CT or only understand CT as a way of problem solving but do not recognise its concepts, sub-dimensions and the mechanism of how it supports content learning (Killen et al., 2023; Sabo et al., 2022). While the different definitions and sub-dimensional variety of CT make it difficult for teachers to understand and integrate CT into their classrooms, researchers suggested that teachers should be educated about the essential concepts of CT in relation to their curriculum that can be used to integrate CT into their teaching through a community-centred approach, including face-to-face and virtual sections (Barr & Stephenson, 2011; Haines et al., 2019; Kite & Park, 2023).

Having a knowledge base for integrating CT into STEM teaching does not mean that STEM teachers will automatically use it in their teaching. In order to transfer the knowledge base to STEM teaching, teachers must first believe that they can integrate CT into their teaching (self-efficacy beliefs) and that integrating CT into their teaching will have a valuable impact on their teaching and their students' learning outcomes (Peters-Burton et al., 2023). In other words, teachers' self-efficacy in relation to CT and their beliefs about the contributions of CT are important for them to integrate CT into their teaching. Researchers have shown that teachers' self-efficacy for CT can be increased through targeted professional development programmes (Rich et al., 2021). Furthermore, the provision of co-design processes, collaboration, regular feedback and iterative classroom practices are shown to be effective in improving teachers' efficacy beliefs, perceptions and willingness to integrate CT into their teaching (Kite & Park, 2023; Peters-Burton et al., 2023; Wu et al., 2020). In addition, other contextual and school-related factors are also important in supporting teachers' integration of CT into their teaching. Jocius et al. (2021) reported that lack of technological tools, overcrowded curriculum, lack of administrative support and low student readiness were reported by teachers as barriers to integrating CT into their lessons. Therefore, these limiting factors also need to be addressed and teachers should be provided with a more positive context to support the integration of CT in different subject areas.

Conclusion

CT is an important concept and skill that all students should develop in order to be informed citizens and equipped for future jobs. In addition to promoting CT in computing related courses, integrating CT into other disciplines such as STEM allows students to see real practices and benefits of CT in a real-world context. On the other hand, the integration of CT into STEM should be well planned in order to get the maximum benefit from the integration. The conclusion section is organised under the two headings as a summary of key points and implications for future practice and research.

Summary of Key Points

Based on the results of the literature review, some key points emerge as follows.

- Three approaches, namely programming activities, problem-solving scenarios and how professionals use CT, are used to promote the integration of CT in STEM.
- Student-centred teaching strategies such as experiential learning, problem-based learning, game-based learning, project-based learning and DBL are shown to be effective in integrating CT in STEM.

- Tools such as Scratch and LEGO used in plugged activities and puzzles and games used in unplugged activities are suggested in the literature as tools for integrating CT in STEM.
- The use of performance assessment, collaborative assessment and self-assessment tools are more appropriate and aligned with CT-integrated STEM learning.
- Teachers' knowledge base, beliefs, self-efficacy and willingness to practice CT in their subject areas emerge as important concepts for implementation.
- The role of collaboration, co-design process and ongoing support from experts are highlighted as crucial for the successful professional development of teachers to support the transition to CT in STEM.

Implications for Future Practice and Research

Based on the review of the literature on the integration of CT and STEM, the chapter draws the following implications. In order to support the integration of CT concepts into STEM, teachers need to be prepared. Thus, future research should focus on finding effective ways to support STEM teachers' practice of CT, as simply waiting for teachers to integrate CT into their teaching is not an easy task for many teachers. Furthermore, while teacher-related factors are essential for integrating CT concepts into STEM, external factors such as a supportive school climate, available infrastructure and guidance are also essential. Thus, educators need to find ways to provide teachers with a suitable environment for integrating CT into STEM.

In terms of research, it is important to compare the effectiveness of different approaches, teaching strategies and assessments to illustrate the useful ways in which teachers can use them in their teaching. For example, while DBL is often used to support CT concepts and STEM explicitly, more studies are needed to show the effectiveness of DBL for integrating CT into STEM. In addition, in-service and pre-service training of STEM teachers is essential to support teachers' perceptions, knowledge and practices of CT concepts. Thus, future researches need to focus on finding effective ways to provide preparation and professional development programmes for STEM teachers.

References

Aksit, O., & Wiebe, E. N. (2020). Exploring force and motion concepts in middle grades using computational modeling: A classroom intervention study. *Journal of Science Education and Technology, 29*(1), 65–82.

Aljowaed, M., & Alebaikan, R. A. (2018). Training needs for computer teachers to use and teach computational thinking skills. *International Journal for Research in Education, 42*(3), 237–284.

Aydeniz, M. (2018). Integrating computational thinking in school curriculum. In M. S. Khine (Ed.), *Computational thinking in the STEM disciplines: Foundations and research highlights* (pp. 253–277). Springer International Publishing. https://doi.org/10.1007/978-3-319-93566-9_13

Barr, V., & Stephenson, C. (2011). Bringing computational thinking to K-12: What is involved and what is the role of the computer science education community? *ACM Inroads, 2*(1), 48–54.

Barth-Cohen, L., Montoya, B., & Shen, J. (2019). Walk like a robot. *Science Scope, 42*(9), 12–17.

Basu, S., Dickes, A., Kinnebrew, J. S., Sengupta, P., & Biswas, G. (2013, May). CTSiM: A computational thinking environment for learning science through simulation and modeling. In *International conference on computer supported education* (Vol. 2, pp. 369–378). SciTePress.

Basu, S., Biswas, G., & Kinnebrew, J. S. (2017). Learner modeling for adaptive scaffolding in a computational thinking-based science learning environment. *User Modeling and User-Adapted Interaction, 27*(1), 5–53.

Beheshti, E., Weintrop, D., Swanson. H., Orton, K., Horn, M. S., Jona, K., & Wilensky, U. (2017). *Computational thinking in practice: How STEM professionals use CT in their work.* Paper presented at the 2017 American Education Research Association Annual Meeting. https://par.nsf.gov/biblio/10026245

Benton, L., Hoyles, C., Kalas, I., & Noss, R. (2017). Bridging primary programming and mathematics: Some findings of design research in England. *Digital Experiences in Mathematics Education, 3*(2), 115–138. https://doi.org/10.1007/s40751-017-0028-x

Bers, M. U., Flannery, L., Kazakoff, E. R., & Sullivan, A. (2014). Computational thinking and tinkering: Exploration of an early childhood robotics curriculum. *Computers & Education, 72*, 145–157.

Black, P., & Wiliam, D. (2009). Developing the theory of formative assessment. *Educational Assessment, Evaluation and Accountability, 21*, 5–31.

Bortz, W. W., Gautam, A., Tatar, D., & Lipscomb, K. (2020). Missing in measurement: Why identifying learning in integrated domains is so hard. *Journal of Science Education and Technology, 29*, 121–136.

Bounou, A., Lavidas, K., Komis, V., Papadakis, S., & Manoli, P. (2023). Correlation between high school students' computational thinking and their performance in STEM and Language courses. *Education Sciences, 13*(11), 1101.

Brennan, K., & Resnick, M. (2012). New frameworks for studying and assessing the development of computational thinking. In *Proceedings of the 2012 annual meeting of the American educational research association, Vancouver, Canada.*

Burgett, T., Folk, R., Fulton, J., Peel, A., Pontelli, E., & Szczepanski, V. (2015). DISSECT: Analysis of pedagogical techniques to integrate computational thinking into K-12 curricula. In *Proceedings—Frontiers in education conference, FIE, 2014.* https://doi.org/10.1109/FIE.2015.7344241

Byrne, E. M., Jensen, H., Thomsen, B. S., & Ramchandani, P. G. (2023). Educational interventions involving physical manipulatives for improving children's learning and development: A scoping review. *Review of Education, 11*(2), e3400.

Cabrera, L., Ketelhut, D. J., Mills, K., Killen, H., Coenraad, M., Byrne, V. L., & Plane, J. D. (2023). Designing a framework for teachers' integration of computational thinking into elementary science. *Journal of Research in Science Teaching.* https://doi.org/10.1002/tea.21888

Caeli, E. N., & Yadav, A. (2020). Unplugged approaches to computational thinking: A historical perspective. *TechTrends, 64*(1), 29–36.

Chai, C. S., Jong, M., Yin, H. B., Chen, M., & Zhou, W. (2019). Validating and modelling teachers' technological pedagogical content knowledge for integrative science, technology, engineering and mathematics education. *Journal of Educational Technology & Society, 22*(3), 61–73.

Chen, P., Yang, D., Metwally, A. H. S., Lavonen, J., & Wang, X. (2023). Fostering computational thinking through unplugged activities: A systematic literature review and meta-analysis. *International Journal of STEM Education, 10*(1), 47.

Chernenkii, A. V. (2020, April). *Using of the aerospace modeling simulator in the educational process.* Paper presented in 2020 V International Conference on Information Technologies in Engineering Education (Inforino) (pp. 1–5). IEEE.

Ching, Y. H., Hsu, Y. C., & Baldwin, S. (2018). Developing computational thinking with educational technologies for young learners. *TechTrends, 62*, 563–573.

Çiftçi, A., & Topcu, M. S. (2023). Improving early childhood pre-service teachers' computational thinking teaching self-efficacy beliefs in a STEM course. *Research in Science & Technological Education, 41*(4), 1215–1241.

Çiftçi, A., & Topçu, M. S. (2022). Improving early childhood pre-service teachers' computational thinking skills through the unplugged computational thinking integrated STEM approach. *Thinking Skills and Creativity,* 101337.

Dabholkar, S., Peel, A., Hao, D., Kelter, J., Horn, M., & Wilensky, U. (2021). Analysis of co-designed biology units integrated with computational thinking activities. In E. de Vries, Y. Hod, & J. Ahn (Eds.), *Proceedings of the 15th international conference of the learning sciences—ICLS 2021* (pp. 665–668). International Society of the Learning Sciences.

Delen, I., & Sen, S. (2023). Effect of design-based learning on achievement in K-12 education: Ameta-analysis. *Journal of Research in Science Teaching, 60*(2), 330–356.

Denning, P. J., & Tedre, M. (2019). *Computational thinking.* MIT Press.

Doppelt, Y., & Schunn, C. D. (2008). Identifying students' perceptions of the important of a design-based learning environment. *Learning Environments Research, 11*(3), 195–209.

Feldhausen, R., Weese, J. L., & Bean, N. H. (2018). Increasing student self-efficacy in computational thinking via STEM outreach programs. In *Proceedings of the 49th ACM technical symposium on computer science education—SIGCSE '18, Baltimore, Maryland, USA.*

Gomes, A., & Mendes, A. J. N. (2007). Learning to program-difficulties and solutions. In *International conference on engineering education* (pp. 1–5).

Govender, I. (2022). Research status in computational thinking in STEM education. In *Advances in research in STEM education.* IntechOpen.

Grover, S., & Pea, R. (2013). Computational thinking in K–12: A review of the state of the field. *Educational Researcher, 42*(1), 38–43. https://doi.org/10.3102/0013189X12463051

Grover, S., & Pea, R. (2018). Computational thinking: A competency whose time has come. In *Computer science education: perspectives on teaching and learning in school* (pp. 19–38).

Gulbahar, Y., Kalelioglu, F., & Kert, S. B. (2018, October). Teaching computational thinking to in-service computer science teachers through a massive open online course. In *E-Learn: World conference on e-learning in corporate, government, healthcare, and higher education* (pp. 922–928). Association for the Advancement of Computing in Education (AACE).

Haines, S., Krach, M., Pustaka, A., Li, Q., & Richman, L. (2019). The effects of computational thinking professional development on STEM teachers' perceptions and pedagogical practices. *Athens Journal of Sciences, 6*, 97–122. https://doi.org/10.30958/ajs

Hava, K., & Koyunlu Ünlü, Z. (2021). Investigation of the relationship between middle school students' computational thinking skills and their STEM career interest and attitudes toward inquiry. *Journal of Science Education and Technology, 30*(4), 484–495.

Hong, Y. C., Qian, Y., & Yang, Y. F. (2021). Examining trajectories of elementary students' computational thinking development through collaborative problem-solving process in a STEM-integrated robotics program. *Journal of Educational Technology Development and Exchange (JETDE), 14*(1), 2.

Hoover, A. K., Barnes, J., Fatehi, B., Moreno-Leon, J., Puttick, G., Tucker-Raymond, E., & Harteveld, C. (2016). Assessing computational thinking in students' game designs. In *Proceedings of the 2016 annual symposium on computer-human interaction in play companion extended abstracts, Austin, Texas, USA.*

Hsu, T.-C., Chang, S.-C., & Hung, Y.-T. (2018). How to learn and how to teach computational thinking: Suggestions based on a review of the literature. *Computers & Education, 126*, 296–310.

Hutchins, N. M., Biswas, G., Maróti, M., et al. (2020). C2STEM: A system for synergistic learning of physics and computational thinking. *Journal of Science Education and Technology, 29*, 83–100. https://doi.org/10.1007/s10956-019-09804-9

ISTE, & CSTA. (2011). *Operational definition of computational thinking for K–12 education.* https://id.iste.org/docs/ct-documents/computational-thinking-operational-definition-flyer.pdf?sfvrsn=2

Israel, M., & Lash, T. (2020). From classroom lessons to exploratory learning progressions: Mathematics + computational thinking. *Interactive Learning Environments, 28*(3), 362–382.

Jocius, R., O'Byrne, W. I., Albert, J., Joshi, D., Robinson, R., & Andrews, A. (2021). Infusing computational thinking into STEM teaching: From professional development to classroom practice. *Educational Technology & Society, 24*(4), 166–179.

Kafai, Y. B., Lee, E., Searle, K., Fields, D., Kaplan, E., & Lui, D. (2014). A crafts-oriented approach to computing in high school: Introducing computational concepts, practices, and perspectives with electronic textiles. *ACM Transactions on Computing Education (TOCE), 14*(1), 1–20.

Kampylis, P., Dagienė, V., Bocconi, S., Chioccariello, A., Engelhardt, K., Stupurienė, G., & Earp, J. (2023). Integrating computational thinking into primary and lower secondary education. *Educational Technology & Society, 26*(2), 99–117.

Killen, H., Coenraad, M., Byrne, V., Cabrera, L., Mills, K., Ketelhut, D. J., & Plane, J. D. (2023). Teacher education to integrate computational thinking into elementary science: A design-based research study. *ACM Transactions on Computing Education, 23*(4), 1–36.

Kirçali, A. Ç., & Özdener, N. (2023). A comparison of plugged and unplugged tools in teaching algorithms at the K-12 level for computational thinking skills. *Technology, Knowledge and Learning, 28*(4), 1485–1513.

Kite, V., Park, S., & Wiebe, E. (2021). The code-centric nature of computational thinking education: A review of trends and issues in computational thinking education research. *SAGE Open, 11*(2), 21582440211016416.

Kite, V., & Park, S. (2023). What's computational thinking?: Secondary science teachers' conceptualizations of computational thinking (CT) and perceived barriers to CT integration. *Journal of Science Teacher Education, 34*(4), 391–414.

Küçükkara, M. F., & Aksüt, P. (2021). An Example of unplugged coding education in preschool period: Activity-based algorithm for problem solving skills. *Journal of Inquiry Based Activities, 11*(2), 81–91.

Ladachart, L., Radchanet, V., Phornprasert, W., & Phothong, W. (2024). Eighth-grade students' content learning and STEM identity as a result of design-based learning. *International Journal of Educational Reform, 19*(2), 300–325. https://doi.org/10.1080/1554480X.2023.2230961

Lee, I., Grover, S., Martin, F., Pillai, S., & Malyn-Smith, J. (2020). Computational thinking from a disciplinary perspective: Integrating computational thinking in K-12 science, technology, engineering, and mathematics education. *Journal of Science Education and Technology, 29*(1), 1–8. https://doi.org/10.1007/s10956-019-09803-w

Lee, V. R., & Vincent, H. (2019). An expansively-framed unplugged weaving sequence intended to bear computational fruit of the loom. In *Proceedings of FabLearn 2019* (pp. 124–127).

Lewis, C. M., & Shah, N. (2012, February). Building upon and enriching grade four mathematics standards with programming curriculum. In *Proceedings of the 43rd ACM technical symposium on computer science education* (pp. 57–62).

Li, X., Xie, K., Vongkulluksn, V., Stein, D., & Zhang, Y. (2021). Developing and testing a design-based learning approach to enhance elementary students' self-perceived computational thinking. *Journal of Research on Technology in Education, 55*(2), 344–368. https://doi.org/10.1080/15391523.2021.1962453

Li, Y., Schoenfeld, A. H., diSessa, A. A., Graesser, A. C., Benson, L. C., English, L. D., & Duschl, R. A. (2020). On computational thinking and STEM education. *Journal for STEM Education Research, 3*(2), 147–166. https://doi.org/10.1007/s41979-020-00044-w

Liao, C. H., Chiang, C. T., Chen, I. C., & Parker, K. R. (2022). Exploring the relationship between computational thinking and learning satisfaction for non-STEM college students. *International Journal of Educational Technology in Higher Education, 19*(1), 43.

Liu, X., Wang, X., Xu, K., & Hu, X. (2023). Effect of reverse engineering pedagogy on primary school students' computational thinking skills in STEM learning activities. *Journal of Intelligence., 11*(2), 36. https://doi.org/10.3390/jintelligence11020036

Looi, C. K., How, M. L., Longkai, W., Seow, P., & Liu, L. (2018). Analysis of linkages between an unplugged activity and the development of computational thinking. *Computer Science Education, 28*(3), 255–279.

Lv, L., Zhong, B., & Liu, X. (2023). A literature review on the empirical studies of the integration of mathematics and computational thinking. *Education and Information Technologies, 28*(7), 8171–8193.

Lytle, N., Cateté, V., Boulden, D., Dong, Y., Houchins, J., Milliken, A., Isvik, A., Bounajim, D., Wiebe, E. N., & Barnes, T. (2019). Use, modify, create: Comparing computational thinking lesson progressions for STEM classes. *ITiCSE*, 395–401.

Malyn-Smith, J, Lee, I. A., Martin, F., Grover, S., Evans, & Pillai, S. (2018) *Developing a framework for computational thinking from a disciplinary perspective.* In S. C. Kong, D. Andone, G. Biswas, T. Crick, H. U. Hoppe, T. C. Hsu, R. H. Huang, K. Y. Li, C. K. Looi, M. Milrad, J. Sheldon, J. L. Shih, K. F. Sin, M. Tissenbaum, & J. Vahrenhold (Eds.), *Proceedings of the international conference on computational thinking education 2018*. The Education University of Hong Kong.

Matsumoto, P. S., & Cao, J. (2017). The development of computational thinking in a high school chemistry course. *Journal of Chemical Education, 94*(9), 1217–1224.

Mendoza Diaz, N. V., Trytten, D. A., Meier, R., & Moore, J. M. (2022). The impact of prior programming experience on computational thinking in first-year engineering experience. In *2022 CoNECD (Collaborative Network for Engineering & Computing Diversity).*

Mensan, T., Osman, K., & Majid, N. A. A. (2020). Development and validation of unplugged activity of computational thinking in science module to integrate computational thinking in primary science education. *Science Education International, 31*(2), 142–149.

Miller, J. (2019). STEM education in the primary years to support mathematical thinking: Using coding to identify mathematical structures and patterns. *ZDM, 51*(6), 915–927.

Mishra, P., & Koehler, M. J. (2006). Technological pedagogical content knowledge: A framework for teacher knowledge. *Teachers College Record, 108*(6), 1017–1054.

Montes-León, H., Hijón-Neira, R., Pérez-Marín, D., & Montes-León, S. R. (2020). Improving computational thinking in secondary students with unplugged tasks. *Education in the Knowledge Society (EKS), 21*, 12–12.

National Research Council. (2012). *A framework for K-12 science education: Practices, crosscutting concepts, and core ideas.* National Academies Press.

Nugent, G., Barker, B., Lester, H., Grandgenett, N., & Valentine, D. (2019). Wearable textiles to support student STEM learning and attitudes. *Journal of Science Education and Technology, 28*, 470–479.

Ogegbo, A. A., & Ramnarain, U. (2022). A systematic review of computational thinking in science classrooms. *Studies in Science Education, 58*(2), 203–230.

Ouyang, Y., Hayden, K. L., & Remold, J. (2018, February). Introducing computational thinking through non-programming science activities. In *Proceedings of the 49th ACM technical symposium on computer science education* (pp. 308–313).

Ozdinç, F., Kaya, G., Mumcu, F., & Yıldız, B. (2022). Integration of computational thinking into STEM activities: An example of an interdisciplinary unplugged programming activity. *Science Activities, 59*(3), 151–159.

Ozyurt, O., & Ozyurt, H. (2023). A large-scale study based on topic modeling to determine the research interests and trends on computational thinking. *Education and Information Technologies, 28*(3), 3557–3579.

Papadakis, S. (2020). Evaluating a teaching intervention for teaching STEM and programming concepts through the creation of a weather-forecast app for smart mobile devices. In M. Kalogiannakis, & S. Papadakis (Eds.), *Handbook of research on tools for teaching computational thinking in P-12 education* (pp. 31–53). IGI Global. https://doi.org/10.4018/978-1-7998-4576-8.ch002

Park, Y.-S., & Park, M. (2018). Exploring students competencies to be creative problem solvers with computational thinking practices. *Journal of the Korean Earth Science Society, 39*(4), 388–400. https://doi.org/10.5467/JKESS.2018.39.4.388

Peel, A., Sadler, T. D., & Friedrichsen, P. (2019). Learning natural selection through computational thinking: Unplugged design of algorithmic explanations. *Journal of Research in Science Teaching, 56*(7), 983–1007.

Peel, A., Dabholkar, S., Anton, G., Wu, S., Wilensky, U., & Horn, M. (2020). A case study of teacher professional growth through co-design and implementation of computationally enriched biology units. In M. Gresalfi, & I. S. Horn (Eds.), *The interdisciplinarity of the learning sciences, 14th international conference of the learning sciences (ICLS) 2020* (Vol. 4, pp. 1950–1957). International Society of the Learning Sciences. https://repository.isls.org//handle/1/6478

Pei, C., Weintrop, D., & Wilensky, U. (2018). Cultivating computational thinking practices and mathematical habits of mind in lattice land. *Mathematical Thinking and Learning, 20*(1), 75–89.

Penuel, W. R., Turner, M. L., Jacobs, J. K., Van Horne, K., & Sumner, T. (2019). Developing tasks to assess phenomenon-based science learning: Challenges and lessons learned from building proximal transfer tasks. *Science Education, 103*(6), 1367–1395.

Peters-Burton, E., Rich, P. J., Kitsantas, A., Stehle, S. M., & Laclede, L. (2023). High school biology teachers' integration of computational thinking into data practices to support student investigations. *Journal of Research in Science Teaching, 60*(6), 1353–1384.

Pietros, J., Sweetman, S., & Shim, M. (2022, December). Teacher practices for computational thinking implementation in elementary science. In *2022 International conference on computational science and computational intelligence (CSCI)* (pp. 2059–2065). IEEE.

Poulakis, E., & Politis, P. (2021). Computational thinking assessment: literature review. *Research on e-learning and ICT in education: Technological, pedagogical and instructional perspectives* (pp. 111–128).

Reiser, M., Binder, M., & Weitzel, H. (2024). Effects of design-based learning arrangements in cross-domain, integrated stem lessons on the intrinsic motivation of lower secondary pupils. *Education in Science, 14*, 1–14. https://doi.org/10.3390/educsci14060607

Rich, P. J., Bartholomew, S., Daniel, D., Dinsmoor, K., Neilsen, M., Reynolds, C., Swanson, M., Winward, E., & Yauney, J. (2021). Trends in tools used to teach computational thinking through elementary coding. *Journal of Research on Technology in Education*, 1–22. https://doi.org/10.1080/15391523.2022.212134

Richardo, R., Dwiningrum, S. I. A., Wijaya, A., Retnawati, H., Wahyudi, A., Sholihah, D. A., & Hidayah, K. N. (2023). The impact of STEM attitudes and computational thinking on 21st-century via structural equation modelling. *International Journal of Evaluation and Research in Education, 12*(2), 571–578.

Rodriguez, B., Kennicutt, S., Rader, C., & Camp, T. (2017, March). Assessing computational thinking in CS unplugged activities. In *Proceedings of the 2017 ACM SIGCSE technical symposium on computer science education* (pp. 501–506).

Rodríguez-Martínez, J. A., González-Calero, J. A., & Sáez-López, J. M. (2020). Computational thinking and mathematics using Scratch: An experiment with sixth-grade students. *Interactive Learning Environments, 28*(3), 316–327.

Romainor, N., Talib, C. A., & Hakim, N. W. A. (2018). The necessity of computational thinking in STEM education: An analysis with recommended research. *SEAMO Recsam, 13*, 1–12.

Saad, A., & Zainudin, S. (2022). A review of project-based learning (PBL) and computational thinking (CT) in teaching and learning. *Learning and Motivation, 78*, 101802.

Sabo, H. C., Odden, T. O. B., & Gregers, T. F. (2022, September). Challenges of preparing secondary STEM pre-service teachers in computational thinking. In *Proceedings of the physics education research conference* (PERC) (pp. 395–400).

Sands, P., Yadav, A., & Good, J. (2018). Computational thinking in K-12: In-service teacher perceptions of computational thinking. In *Computational thinking in the STEM disciplines: Foundations and research highlights* (pp. 151–164).

Saritepeci, M. (2020). Developing computational thinking skills of high school students: Design-based learning activities and programming tasks. *The Asia-Pacific Education Researcher, 29*(1), 35–54.

Sen, C., Ay, Z. S., & Kiray, S. A. (2021). Computational thinking skills of gifted and talented students in integrated STEM activities based on the engineering design process: The case of robotics and 3D robot modeling. *Thinking Skills and Creativity, 42*, 100931.
Sun, L., You, X., & Zhou, D. (2023). Evaluation and development of STEAM teachers' computational thinking skills: Analysis of multiple influential factors. *Education and Information Technologies*, 1–35.
Sun, D., Looi, C. K., Li, Y., Zhu, C., Zhu, C., & Cheng, M. (2024). Block-based versus text-based programming: a comparison of learners' programming behaviors, computational thinking skills and attitudes toward programming. *Educational Technology Research and Development*. https://doi.org/10.1007/s11423-023-10328-8
Sung, W., Ahn, J., & Black, J. B. (2017). Introducing computational thinking to young learners: Practicing computational perspectives through embodiment in mathematics education. *Technology, Knowledge and Learning, 22*(3), 443–463.
Swaid, S. I. (2015). Bringing computational thinking to STEM education. *Procedia Manufacturing, 3*, 3657–3662.
Taylor, K., & Baek, Y. (2019). Grouping matters in computational robotic activities. *Computers in Human Behavior, 93*, 99–105.
Tran, Y. (2019). Computational thinking equity in elementary classrooms: What third-grade students know and can do. *Journal of Educational Computing Research, 57*(1), 3–31.
Urban, M. J. (2015). *Improving K-12 STEM education outcomes through technological integration*. IGI Global.
Valentine, K. D. (2018). Tinkering with Logo in an elementary mathematics methods course. *Interdisciplinary Journal of Problem-Based Learning, 12*(2).
Verawati, N. N. S. P., Rijal, K., & Grendis, N. W. B. (2023). Examining STEM students' computational thinking skills through interactive practicum utilizing technology. *International Journal of Essential Competencies in Education, 2*(1), 54–65.
Wang, C., Shen, J., & Chao, J. (2022). Integrating computational thinking in STEM education: A literature review. *International Journal of Science and Mathematics Education, 20*, 1949–1972. https://doi.org/10.1007/s10763-021-10227-5
Weintrop, D., & Wilensky, U. (2013). Supporting computational expression: How novices use programming primitives in achieving a computational goal. *American Education Researchers Association, San Francisco, CA, USA*.
Weintrop, D., Beheshti, E., Horn, M., Orton, K., Jona, K., Trouille, L., & Wilensky, U. (2016). Defining computational thinking for mathematics and science classrooms. *Journal of Science Education and Technology, 25*(1), 127–147. https://doi.org/10.1007/s10956-015-9581-5
Wing, J. M. (2006). Computational thinking. *Communications of the ACM, 49*(3), 33–35.
Wing, J. M. (2008). Computational thinking and thinking about computing. *Philosophical Transactions of the Royal Society of London A: Mathematical, Physical and Engineering Sciences, 366*(1881), 3717–3725.
Wu, S., Peel, A., Bain, C., Anton, G., Horn, M., & Wilensky, U. (2020). Workshops and codesign can help teachers integrate computational thinking into their K-12 STEM classes. In *Proceedings of international conference on computational thinking education 2020*.
Yadav, A., Krist, C., Good, J., & Caeli, E. N. (2018). Computational thinking in elementary classrooms: Measuring teacher understanding of computational ideas for teaching science. *Computer Science Education, 28*(4), 371–400.
Yang, D., Baek, Y., Ching, Y.-H., Swanson, S., Chittoori, B., & Wang, S. (2021). Infusing computational thinking in an integrated STEM curriculum: User reactions and lessons learned. *European Journal of STEM Education, 6*(1), 1–14. https://doi.org/10.20897/ejsteme/9560
Yeni, S., Grgurina, N., Saeli, M., Hermans, F., Tolboom, J., & Barendsen, E. (2024). Interdisciplinary integration of computational thinking in K-12 education: A systematic review. *Informatics in Education*. https://doi.org/10.15388/infedu.2024.08
Yu, X., & Guo, X. (2018). Case study on "STEM+computational thinking" education model in Chinese K-12 schools. *Journal of Science Education, 19*(1), 163–177.

Zhang, Y., Luo, R., Zhu, Y., & Yin, Y. (2021). Educational robots improve K-12 students' computational thinking and STEM attitudes: Systematic review. *Journal of Educational Computing Research, 59*(7), 1450–1481.

Zhong, B., Wang, Q., Chen, J., & Li, Y. (2016). An exploration of three-dimensional integrated assessment for computational thinking. *Journal of Educational Computing Research, 53*(4), 562–590.

Chapter 11
Investigating the Impact of Physical Programming on Students' Attitudes Towards Physics: A Design-Based Learning Approach

Effransia Tzagkaraki, Stamatios Papadakis, and Michail Kalogiannakis

Abstract Physical programming has frequently found application within the scholastic setting, wherein students actively engage in authentic exercises to discern resolutions for tangible predicaments through direct observation and empirical trials. In our specific context, physical programming was employed to teach electrical circuits to fifth-grade students within an elementary school environment. The overarching objective of this present study is to scrutinize the impact and magnitude of influence exerted by physical programming in conjunction with block-based pedagogical techniques utilizing the Micro:bit board on the attitudes, motivation, and self-efficacy of children concerning the field of physics, combined with the Design-Based Learning approach. Notably, discernible disparities were ascertained within the experimental cohort.

Keywords Physical computing · Physics · Elementary school · Attitudes

Introduction

Using tangible devices, particularly single-board configurations, is a versatile tool that educators and learners can employ innovatively across various educational settings (Kalogiannakis et al., 2021; Vordou & Romero, 2021) and within

---

E. Tzagkaraki
Department of Preschool Education, University of Crete, Crete, Greece

S. Papadakis · M. Kalogiannakis (✉)
Department of Special Education, University of Thessaly, Thessaly, Greece
e-mail: mkalogian@uth.gr

S. Papadakis
e-mail: stpapadakis@uoc.gr

M. Saritepeci and H. Yildiz Durak (eds.), *Integrating Computational Thinking Through Design-Based Learning*, https://doi.org/10.1007/978-981-96-0853-9_11

diverse educational applications that enable students to engage in collaborative and interactive projects (Katterfeldt et al., 2018).

We have already implemented physical programming in numerous educational environments through these devices and boards. This approach offers a substantial advantage, engaging students in the complete creative process, from designing and building to programming and operating robots. This hands-on experience fosters the development of unique creations that spring from their imagination and creativity (Min & Kim, 2021). Notably, physical computing has demonstrated its capacity to augment intrinsic motivation for learning, regardless of students' backgrounds. Furthermore, it promotes initiative and active participation, enhancing self-regulated learning abilities. These elements are deemed pivotal, as they significantly impact the efficacy of the learning process (Przybylla & Romeike, 2018). A Design-Based Learning approach is an appropriate framework for enhancing students' computational thinking and active engagement (Papavlasopoulou et al., 2018).

The study investigates how physical programming and block-based instruction facilitated by the Micro:bit board influence children's attitudes toward physics. In pursuit of this objective, we examine potential disparities associated with teaching intervention by delving into the following research inquiries:

RQ1: Do students' perceptions of the teacher's role differ?
RQ2: Is there a difference in students' perceptions of the importance attributed to Physics for society?
RQ3: Are there changes in student motivation?
RQ4: Do perceptions of self-efficacy differ?

Next, we present results obtained through the questionnaire (pre-test and post-test). Finally, the main findings are discussed for suggestions and future research.

Background

Physical Programming: Conceptualizing and Supporting Learning

The possibilities for harnessing physical devices continue to expand, gaining popularity among students of all ages (Gendreau Chakarov et al., 2021; Sentance et al., 2017). In physical computing, we refer to creating devices equipped with sensors that can detect, respond to, and interact with users. What may have begun as tinkering with tangible devices as a hobby, where natural and digital worlds converge to produce original artefacts delightfully and creatively, has evolved. Today, we actively seek effective strategies to leverage their full potential and seamlessly integrate them into educational curricula (Kalelioglu & Sentance, 2020).

Even novice users, including young learners, can grasp fundamental programming concepts through physical and visual programming exposure. This pedagogical process is engaging, employing blocks and colours. Each block possesses distinct attributes, such as colour and shape, implicitly conveying its semantics and potential for integration with other blocks. Within various command palettes offered by programming environments like Micro: bit, students can craft sequences of commands to achieve desired outcomes in program operation. The most notable advantage of physical programming lies in its capacity to visualize program execution in real-time and physical space (Baweja, 2020; Ntourou et al., 2021).

The impact of physical programming on learning has been examined from multifaceted perspectives. Numerous studies explore its correlation with enhancing computational thinking (Kastner-Hauler et al., 2022; Min & Kim, 2021). A study by Kwon et al. (2021) found that debugging skills significantly influence students' self-efficacy and propensity to engage with computational thinking challenges. Furthermore, such a framework can intersect with diverse cognitive domains and serve as a platform for interdisciplinary pedagogical practices (Vermillion & de Salvatierra, 2019). In alignment with this, research posits that immersion in natural systems actively engages the mind and body, simultaneously stimulating multiple senses and augmenting the learning experience (Neutens, 2022).

Studies have demonstrated that students can cultivate a spectrum of skills through applications of physical programming, including computational thinking, coding, algorithmic reasoning, creativity, collaboration, communication, and more (Kwon et al., 2021). Because problem-solving occurs in real time, students can bridge conceptual content to everyday activities (Theodoropoulos et al., 2018). Additionally, active engagement in learning bolsters motivation and self-efficacy as students embark on the creative journey of envisioning, planning, and managing their creations (Przybylla & Romeike, 2018).

Project—Base Learning and Physical Programming

In their research, Sáez-López et al. (2016) find that combining a visual programming language with a student-centered pedagogical approach improves learning outcomes by strengthening the motivation, enthusiasm and communication skills of those involved.

In this context, we attempted to combine physical programming applications, specifically the Micro:bit microcontroller, with the didactic approach of Project-Based Learning (PBL). Essentially, it is a widespread approach that aims to create situations that encourage creativity by solving real problems (Kwon et al., 2021). Project-based learning relies mainly on student experience and is driven by the project. It needs to be more structured and is an effective way of linking together various lessons and units of a course. In essence, it is extensive research where the students in each phase of the project use the research process, being responsible for their learning through experiential learning. It is an active construction of knowledge where previous knowledge and experiences are connected, reconstructed and given new content through research and more profound understanding (Salam, 2022).

Micro: Bit as Students Targeted Device

Over the past decade, there has been a notable increase in specialized studies, particularly those focused on the BBC Micro:bit single-board microprocessor. The BBC micro: bit represents a low-cost, programmable device designed to foster learning through tangible tools, enhance motivation, and create opportunities for collaboration and creativity (Kalelioglu & Sentance, 2020).

Research concerning students' perspectives on utilizing the Micro:bit predominantly yields positive feedback. Students commonly report finding it engaging, enjoyable, user-friendly, and personally satisfying, albeit with variations based on their prior experience (Kalogiannakis et al., 2021).

Incorporating the Micro: bit presents an avenue for addressing challenges associated with teaching and learning programming and algorithmic thinking. It can boost students' enthusiasm for computer science and catalyze their pursuit of successful careers in STEM fields (Sentance et al., 2017). This device offers a user-friendly and enjoyable introduction to robotics, allowing students to actively participate in the coding and programming process that might otherwise seem complex and inaccessible (Bence, 2023). Notably, this benefit also extends to educators, enabling them to enrich their teaching methods uniquely that were previously unattainable, capitalizing on the device's open and versatile nature.

Attitudes and Students' Attitudes to Physics

It is imperative to elucidate the concept of "attitude," a pivotal focal point in the current study. In this context, attitude pertains to an individual's inclination to categorize objects or events and subsequent reactions and evaluations (Nja et al., 2022). To assess attitudes, evaluative responses evaluate students' favourable or unfavourable inclinations regarding their performance in various activities (Love & Asempapa, 2022). It is noteworthy to consider the conclusion Fishman et al. (2021) drew, which emphasizes the need for greater precision in defining attitudes. Our endeavour seeks to adhere to standardized definitions and validated measurement methodologies to identify predictive factors. Given its intricacy, we consider Intrinsic motivation, Career Motivation, Grade Motivation, and Self-Efficacy in our study to measure attitudes (Astalini et al., 2018; Hillman et al., 2016; Salta & Koulougliotis, 2015; Tai et al., 2022).

When assessing attitudes, it is crucial to recognize that they evolve. Nonetheless, it is a vital process, as the outcomes derived from such evaluations offer insights into crafting appropriate learning strategies (Astalini et al., 2018).

Students' attitudes towards physics lessons vary from positive to negative perspectives. Positive attitudes may be linked to joy, excitement, contentment, or heightened curiosity towards learning activities. Conversely, negative attitudes can be associated with disinterest in attending classes, apathy towards the teacher's explanations, and diminished inspiration and creativity. Considering this, students harbouring positive attitudes towards physics tend to display heightened motivation and engagement. In contrast, those with negative attitudes distaste the subject and the instructor (Astalini et al., 2018).

The research underscores the potential of employing physical computing and tangible devices, given their substantial influence on various aspects of human behaviour, including attitudes, which are integral to effective learning. It has been observed that using physical devices notably impacts students' attitudes, particularly concerning computational thinking (Theodoropoulos et al., 2018). Furthermore, Design-based Learning is a widespread approach that aims to create situations that encourage creativity by solving real-world problems (Kwon et al., 2021). The project has a less specific structure based on the students' experience and guided by the project. It is a particularly effective way of connecting many lessons and other course units. It is extensive research where the students in each project phase use the research process to be responsible for their learning.

Method

Design

The present study applied a quasi-experimental design (Cohen et al., 2007; Kanaki & Kalogiannakis, 2023). The research was implemented in three phases from February to May 2023. In the first phase, students were given a questionnaire with 28 attitude detection questions (Pre-test). In the second phase, the teaching intervention of 10 teaching hours was carried out with the implementation of the planned activities. After the completion of the experimental design, the Post-test was carried out to evaluate the effect of the experimental teaching, according to the purpose of the research.

Sample

The present study's sample was 140 students from four Greek public schools. In each school, there were two different departments. One was chosen by lottery as the experimental group, so the two groups (control and experimental) were equivalent (Table 11.1). All students were born in 2012 (they are in 5th grade) and had the basic knowledge of using visual programming commands. The majority had used Scratch for more than five teaching hours, as the IT teacher informed us.

Table 11.1 Demographic information

Group	Male	Female	N	%
Control	31	37	68	48.6
Experimental	34	38	72	51.4
Total	65	75	140	100.0

Researching Tool

In 2000, Weinburgh and Steele developed an attitude measurement instrument for African American male and female students attending fifth-grade elementary school in an urban area. Their instrument, the Modified Attitudes Toward Science Inventory (mATSI), is a modified and shortened version of the Attitudes Toward Science Inventory: version a (ATSI:a) instrument (Tai et al., 2022).

We searched databases and published articles in reputable journals to find a tool. We studied their content for relevance to the research topic and obtained copies of the instruments to access their questions. From all the researchers found, we came to those of Weinburgh and Steele (2000), Hillman et al. (2016), and Tai et al. (2022).

The mATSI version was the basis of the instrument used for the present research. The items in our survey questionnaire were selected to examine some critical indicators, as seen from the literature. The selection criteria of the tool were the time of publication. They were preferred because they are relatively recent. Another important criterion was the number of references from other researchers, as the frequency of use strengthens the consistency of the tool under consideration. Also, the measurement scales were suitable for the statistical procedures used in the surveys above, enhancing their reliability and validity. Although a search was made for available reviews of the tools, none were found (Table 11.2).

In order to assess the conceptual construct and verify whether the elements—statements that belong to the same category constitute a Common Factor, a Factor Analysis was carried out, and an Exploratory Factor Analysis was applied. Initially, the KMO and Bartlett's Test of Sphericity value was calculated to verify whether the factorial structure can be analyzed and whether there is a correlation between the examined variables. The value of 0.830 shows that we can analyze our data, and in the sphericity test, $p < 0.05$ ($p = 0.000$), so there is a correlation between the variables.

The tailored questionnaire comprising six distinct categories was developed as a research instrument to align with the specific research requirements. Each category encompasses five questions, except for the initial category, which comprises three questions. The designated categories are as follows: (1) Perceptions of the teacher, (2) The societal significance of Physics, (3) Intrinsic motivation, (4) Career aspirations, (5) Self-efficacy, and (6) Academic motivation. In April 2022, a preliminary questionnaire was administered to a cohort of 30 students in a pilot study, leading to subsequent refinements of its final iteration. Particular emphasis was placed on formulating questions conducive to responses from all participating children. Prior to

Table 11.2 Tool selection criteria

Timeliness	Hillman et al. (2016) (N = 549) Tai et al. (2022) (N = 2.016)
References (Last searched February 2023)	Google Scholar: Weinburgh and Steele (2000) for 5th grade students (131 references) Hillman et al. (2016) (*78* references) Tai et al. (2022) (*8* references)
Measurement Scale—Reliability Index	Likert Scale Weinburgh and Steele (2000) *Cronbach's* $\alpha = 0.72$*–*0.78 Hillman et al. (2016) *Cronbach's* $\alpha = 0.909$ Tai et al. (2022) Factor analysis shows that the data support the proposed model and is satisfactory in aligning with the structure of the collected observations
Relevance to the topic	Measure of student's attitudes toward science

the commencement of the questionnaire administration process, essential concepts such as challenges, career prospects, and professional advantages were elucidated to the participants. Moreover, all requisite permissions and approvals were duly obtained to conduct the research (Petousi & Sifaki, 2020).

Data Analysis

The data underwent analysis using the statistical software SPSS version 26. This involved comparing variables and generating corresponding tables and graphs. To assess the reliability of the dataset, Cronbach's α coefficient was computed, yielding a value of 0.879.

Results

This section unveils the study's findings and addresses the research inquiries raised. The non-parametric Wilcoxon test served as the analytical tool to assess the alterations between the pre-test and post-test results, as illustrated in Tables 11.3 and 11.4.

Upon scrutinizing the variations in post-test outcomes within each category and group, it became evident that the experimental group exhibited an increment in agreement percentages for five out of the six categories evaluated, based on the pertinent

Table 11.3 Wilcoxon signed ranks test per category

		N	Mean rank	Sum of ranks
teacher_perception_post-teacher_perception_pre	Negative ranks	43a	47.33	2035.00
	Positive ranks	61b	56.15	3425.00
	Ties	36c		
	Total	140		
Physics_society_post—Physics_society_pre	Negative ranks	59d	63.18	3727.50
	Positive ranks	59e	55.82	3293.50
	Ties	22f		
	Total	140		
Intrinsic_motivation_post—Intrinsic_motivation_pre	Negative ranks	69g	64.62	4459.00
	Positive ranks	50h	53.62	2681.00
	Ties	21i		
	Total	140		
Career_motivation_post—Career_motivation_pre	Negative ranks	66j	61.10	4032.50
	Positive ranks	58k	64.09	3717.50
	Ties	16l		
	Total	140		
Self_efficacy_post—Self_efficacy_pre	Negative ranks	58m	55.84	3238.50
	Positive ranks	56n	59.22	3316.50
	Ties	26o		
	Total	140		
Grade_motivation_post—Grade_motivation_pre	Negative ranks	58p	55.73	3232.50
	Positive ranks	53q	56.29	2983.50
	Ties	29r		
	Total	140		

Table 11.4 Wilcoxon test statistics per category

Test statistics		
	Z	Asymp. Sig. (2-tailed)
teacher_perception_post—teacher_perception_pre	-2.283^{b}	0.022
Physics_society_post—Physics_society_pre	-0.585^{c}	0.558
Intrinsic_motivation_post—Intrinsic_motivation_pre	-2.371^{c}	0.018
Career_motivation_post—Career_motivation_pre	-0.394^{c}	0.693
Physics_society_post—Physics_society_pre	-0.111^{b}	0.912
Grade_motivation_post—Grade_motivation_pre	-0.369^{c}	0.712

[a]Wilcoxon signed ranks test
[b]Based on negative ranks
[c]Based on positive ranks

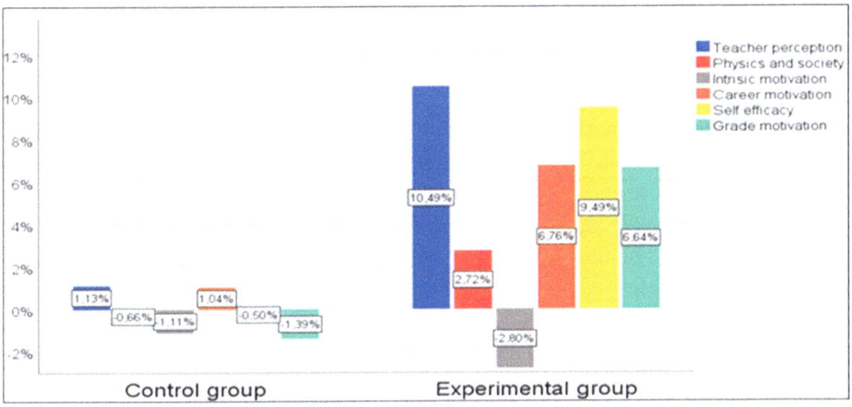

Fig. 11.1 Post-test differences per category and group

statements in the questionnaire. However, a decline in the agreement percentage was observed in the category related to intrinsic motivation, as depicted in Fig. 11.1.

RQ1: Do Students' Perceptions of the Teacher's Role Differ?

Both groups have a general tendency to accept the perception that teachers make physics interesting (Q1), present the material in an interesting way (Q2), and are willing to help individually (Q3). Most students agree or agree with the relevant statements. In the Experimental group, in the post-test, there is a slightly higher percentage of children who "neither agree nor disagree" about their interest in Physics and the way the material is presented (19.44%) compared to the Control Group (14.71%). However, the percentage of those who "strongly agree" that teachers make physics enjoyable is higher in the Experimental group, with 80.56% after, compared to 77.95% in the Control Group.

In the Control Group, there is a higher percentage of students who "agree" that the teacher is willing to help individually (44.1%) compared to the Experimental Group (34.7%). Also, after the intervention in the control group, the percentage of students who "strongly agree" is lower (19.12%) than the experimental group (26.39%). The percentage of children who "strongly disagree" is generally low in both groups for all three questions.

RQ2: Is There a Difference in Students' Perceptions of the Importance Attributed to Physics for Society?

In the Experimental group, they agree more that Physics helps solve everyday problems (62.50% before and 66.66% after agree or strongly agree) than the Control Group (44.12% before and 52.94% after) (Q4) and that Physics is helpful to everyone and not just scientists (70.8% before and 75% after) relative to the control group which maintained its levels of agreement (75% before and after) (Q7). Regarding the statement that physics helps to understand today's world, the Experimental group shows slightly more agreement, with 62.50% before and 65.28% after the intervention, compared to 73.53% before and 58.82% after the Control group (Q5).

RQ3: Are There Changes in Student Motivation?

Within the spectrum of motivational factors, our study delves into intrinsic, career, and grade motivation. Regarding intrinsic motivation, a shift in perception regarding the ease of physics is noteworthy in the post-test for both groups. A closer examination of the detailed statistical data reveals an emerging difference. In the control group, the total percentage of students in agreement remains almost consistent between the pre-test (54.41%) and post-test (55.88%). Conversely, in the experimental group, these percentages demonstrated a notable increase, transitioning from 55.55% in the pre-test to 63.89% in the post-test (Refer to Figs. 11.2 and 11.3).

In the post-test, the experimental group exhibited an overall increase in the percentage of agreement regarding their preference for the physics lesson, rising from 54.1 to 63.9%. Conversely, in the control group, the corresponding percentages decreased from 63.3 to 61.7%. A similar trend is evident when considering the

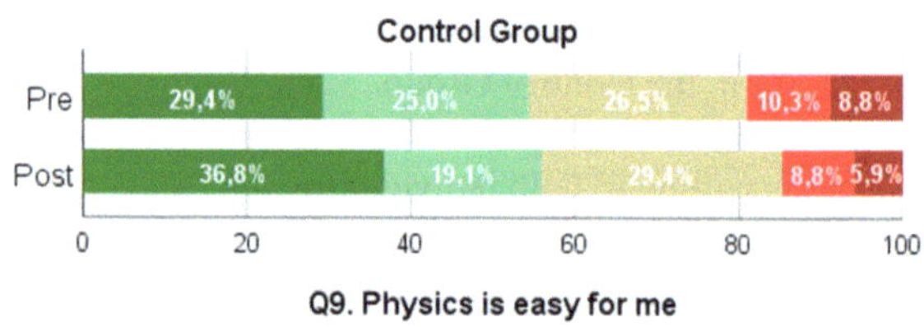

Fig. 11.2 Q9. Physics is easy for me (control group)

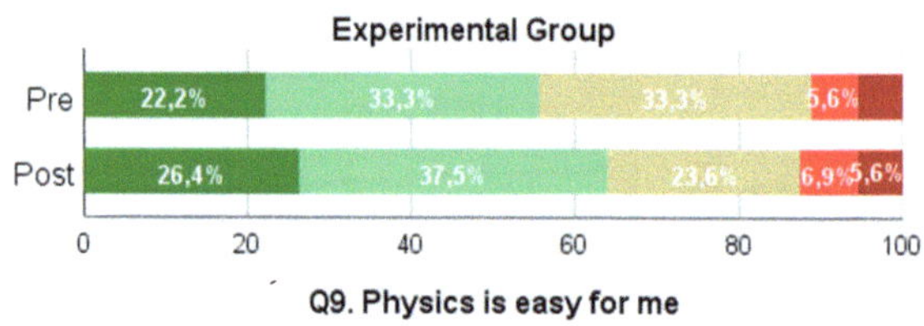

Fig. 11.3 Q9. Physics is easy for me (experimental group)

perceived challenge of physics tasks. The experimental group saw a 5.5% increase in their agreement, while the control group experienced a 2.9% decrease compared to the pre-test.

Turning to career motivation, notable shifts were observed within the experimental group. There was increased agreement among students regarding their interest in pursuing a future physics-related job and its potential career benefits (Figs. 11.4 and 11.5). In the context of their future professions and inclination towards physics-related careers, the experimental group displayed a significant 15% increase in agreement. In contrast, the control group experienced a 19% decline.

Upon analyzing the grade motivation data and juxtaposing questions 26 and 27 (See Figs. 11.6, 11.7, 11.8 and 11.9), a shared concern emerges from both groups regarding their performance in the physics course. They express a need for high grades, particularly in the context of examinations. A more detailed examination of the experimental group reveals a subtle decline in the percentage of students expressing apprehension about their grades in the post-test. This decline signifies a boost in the group's confidence, a trend corroborated by the subsequent statement.

In the concluding statement on the students' willingness to outperform their peers, notable trends become apparent. In the control group, there is a discernible decrease in students' inclination to surpass their classmates' test performances, with a decline

Fig. 11.4 Q14. I want a job related to Physics because I am interested in (control group)

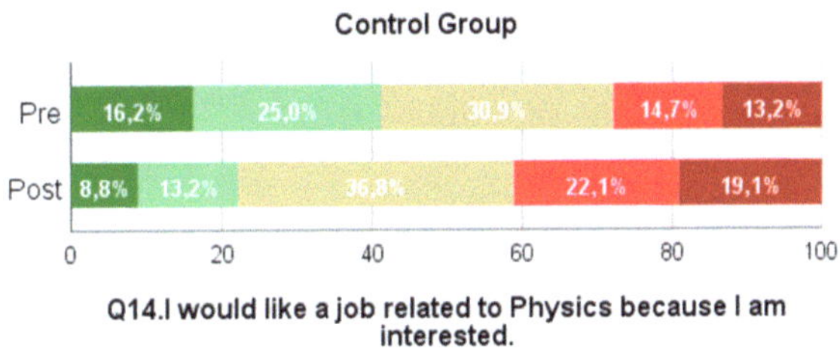

Fig. 11.5 Q14. I want a job related to Physics because I am interested in (experimental group)

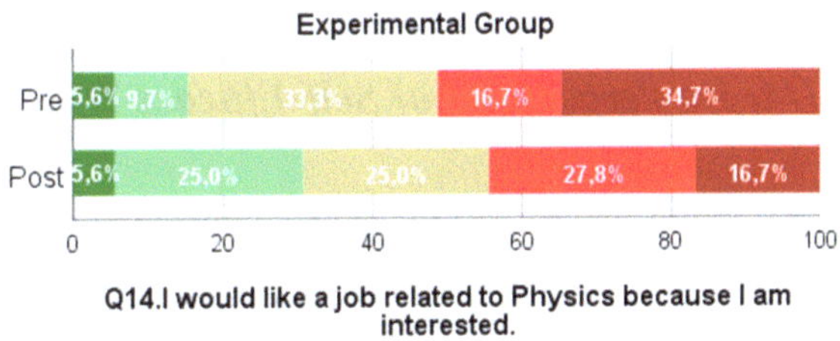

Fig. 11.6 Q26. I am concerned about the grade I will get in Physics (control group)

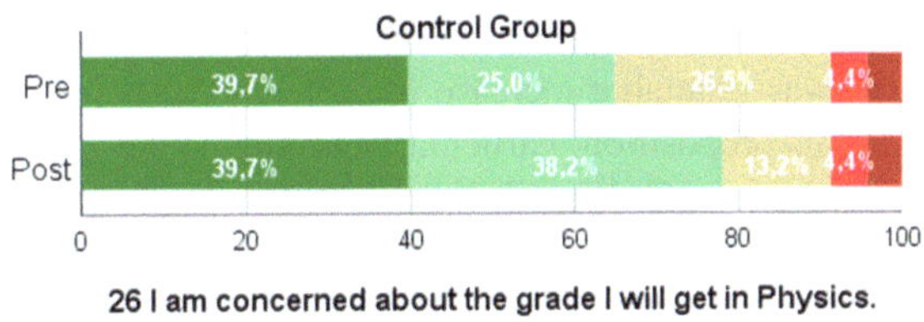

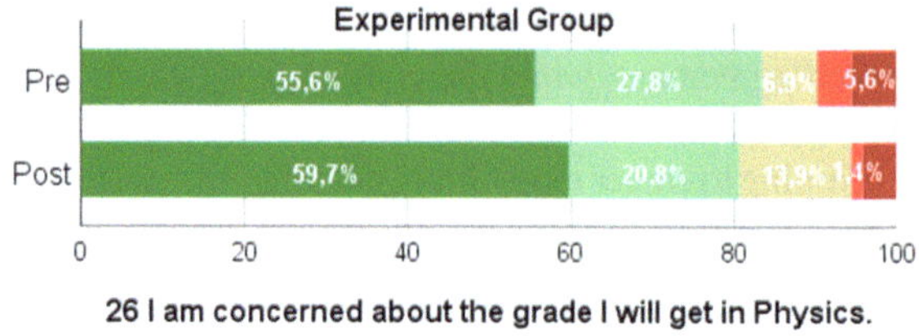

Fig. 11.7 Q26. I am concerned about the grade I will get in Physics (experimental group)

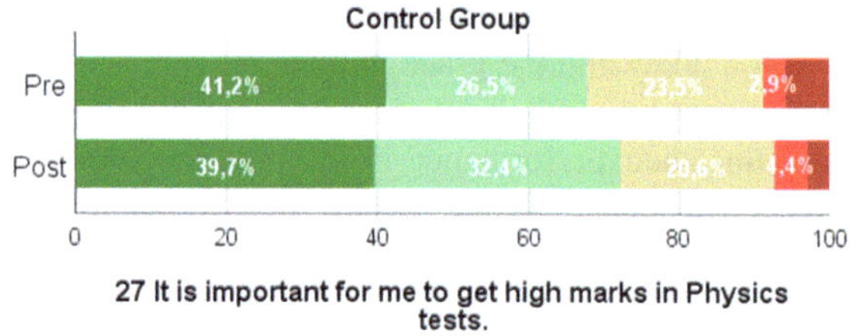

Fig. 11.8 Q27. I need high marks on the Physics tests (control group)

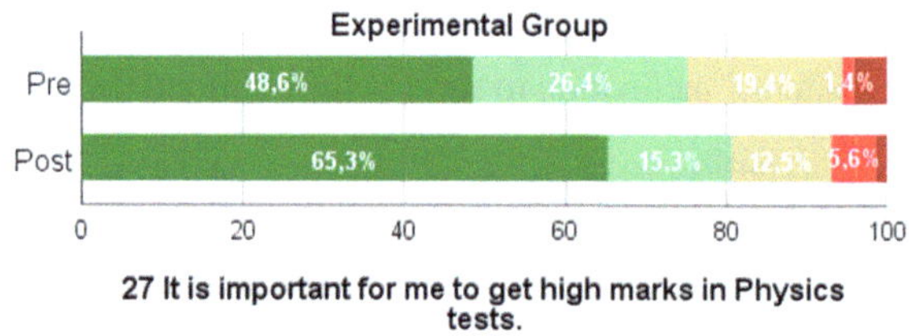

Fig. 11.9 Q27. I need high marks on the Physics tests (experimental group)

of 10.3%. Conversely, within the experimental group, students exhibit a heightened level of motivation in the post-test, evidenced by an 8.4% increase in the percentage of students in agreement with this sentiment.

RQ4: Do Perceptions of Self-Efficacy Differ?

The following question pertains to the student's comprehension of physics; intriguingly, the two groups exhibited contrasting outcomes in the post-test. The control group saw a decrease of 13.3% in their percentage, while the experimental group recorded an increase of 5.6% (Refer to Figs. 11.10 and 11.11).

When considering students' self-confidence, specifically their belief in performing well in physics assignments, noteworthy patterns emerge. In the experimental group, there was a notable increase of 2.8% in the percentage of students in agreement with this proposition. Particularly striking is the 7% rise in students who agree with this statement. Conversely, in the control group, a contrasting trend is evident. The number of students who agreed with the statement decreased by 8.8%, dropping from 63.2 to 54.4%.

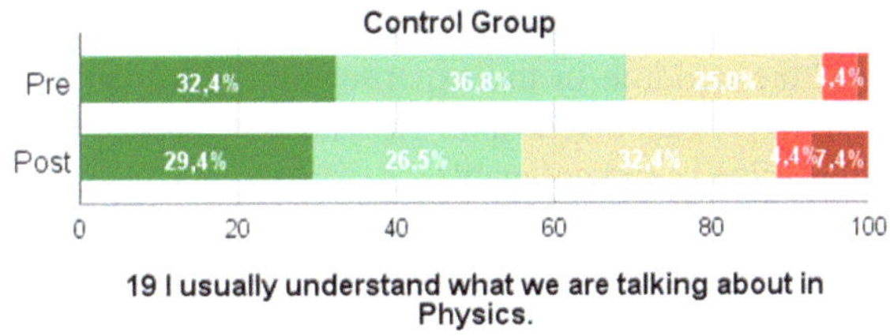

Fig. 11.10 Q19. I understand what we are talking about in Physics (control group)

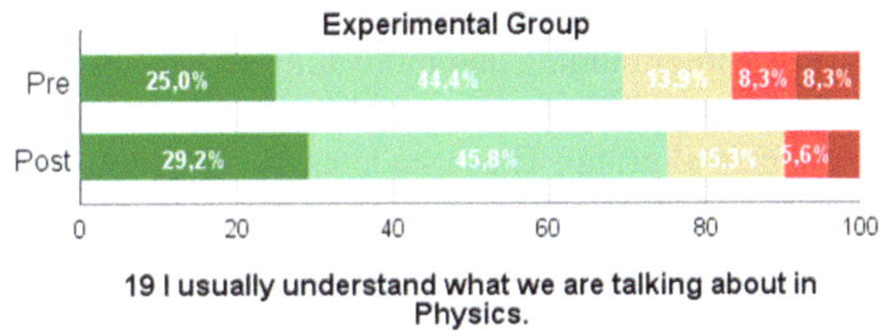

Fig. 11.11 Q19. I understand what we are talking about in Physics (experimental group)

Discussion and Conclusion

In this study, we explore the impact of a proposed method involving the application of physical computing on the attitudes of fifth-grade elementary school students during their physics lessons, with a specific focus on electricity and the utilization of the BBC Micro:bit board. Our data analysis has yielded valuable insights.

Given the literature's absence of a definitive definition for attitudes, we opted to investigate certain facets of attitudes as indicative measures. The primary questions guiding our research were: How does the teacher's approach influence student interest? Does the use of the Micro: bit impact student motivation and self-efficacy?

For the first question, our findings underscore the pivotal role of the teacher, their instructional methods, and the individualized support they offer, which significantly influence students in both groups. Notably, the experimental group showed a remarkable increase in the percentage of agreement (89%) with the corresponding statements. This aligns with existing literature, highlighting that engaging in physical programming, particularly with the Micro:bit, stimulates children's interest due to its inherently appealing teaching approach (Kalogiannakis et al., 2021; Sentance et al., 2017). Furthermore, the experimental group demonstrated a greater need for individual assistance, which is to be expected, given that the proposed approach deviates from their accustomed methods.

The children in the experimental group displayed a heightened recognition of the importance of physics in solving everyday problems compared to the control group. This finding holds substantial significance as our aim, through physical computing and tangible devices, is to bridge children's everyday reality with digital tools (Min & Kim, 2021).

Motivation plays a pivotal role in predicting engagement in physics (Przybylla & Romeike, 2018). To explore this further, we delved into specific motivational aspects, including intrinsic, career, and grade motivation. In the context of intrinsic motivation, the children in the experimental group demonstrated reinforced motivation. They expressed that despite the challenges posed by the activities, they find physics approachable and enjoyable.

Noteworthy is the agreement among children in the experimental group (55% as opposed to 47% in the pre-test) regarding the value of understanding physics-related concepts in their future careers and their desire for related employment opportunities. This finding aligns with prior research by Kalogiannakis, Tzagkaraki, and Papadakis (2021) and Love and Asempapa (2022), highlighting the link between physical programming and motivating students to pursue careers in STEM fields.

Significant differences were also observed in the grade motivation category for the experimental group. These students were keen to perform well in physics exercises and surpass their classmates. The intervention with physical computing markedly enhanced their enthusiasm for these aspects, reflecting a positive impact on self-efficacy (Theodoropoulos et al., 2018).

Our study contributes to the growing body of empirical evidence underscoring the importance of incorporating physical and visual programming into subjects beyond computer science courses. The effects of such interventions may take time to yield dramatic changes. Nevertheless, the positive influence of our intervention on student attitudes bolsters the endeavour to enhance the pedagogical approach in physics, ultimately strengthening computational thinking skills. Conducting longitudinal research would provide a deeper understanding of the topics discussed herein.

References

Astalini, A., Kurniawan, D. A., Perdana, R., & Kurniasari, D. (2018). Identification of student attitudes toward physics learning at Batanghari district high school. *The Educational Review, USA, 2*(9), 475–484. https://doi.org/10.26855/er.2018.09.003

Baweja, A. (2020). Physical programming language. *International Journal of Scientific & Technology Research, 9*(06), 421–424. https://doi.org/10.6084/m9.figshare.13303448.v3

Bence, G. (2023). The introduction of Micro:bit in elementary school, from unplugged activity to programs. *Olympiads in Informatics, 17*, 33–42. https://doi.org/10.15388/ioi.2023.03

Cohen, L., Manion, L., & Morrison, K. (2007). *Research methods in education* (6th ed.). Routledge/Taylor & Francis Group.

Fishman, J., Yang, C., & Mandell, D. (2021). Attitude theory and measurement in implementation science: A secondary review of empirical studies and opportunities for advancement. *Implementation Science, 16*(1), 87. https://doi.org/10.1186/s13012-021-01153-9

Gendreau Chakarov, A., Biddy, Q., Hennessy Elliott, C., & Recker, M. (2021). The data sensor hub (DaSH): a physical computing system to support middle school inquiry science instruction. *Sensors, 21*(18). https://doi.org/10.3390/s21186243

Hillman, S. J., Zeeman, S. I., Tilburg, C. E., et al. (2016). My Attitudes Toward Science (MATS): The development of a multidimensional instrument measuring students' science attitudes. *Learning Environments Research, 19*, 203–219. https://doi.org/10.1007/s10984-016-9205-x

Kalelioglu, F., & Sentance, S. (2020). Teaching with physical computing in school: The case of the Micro:Bit. *Educational Technology & Society, 25*, 2577–2603. https://doi.org/10.1007/s10639-019-10080-8

Kalogiannakis, M., Tzagkaraki, E., & Papadakis, St. (2021). A systematic review of the use of BBC Micro:bit in primary school. In *Proceedings of the 10th virtual edition of the international conference new perspectives in science education*, 18–19 (pp. 379–384). https://doi.org/10.26352/F318_2384-9509

Kanaki, K., & Kalogiannakis, M. (2023). Sample design challenges: An educational research paradigm. *International Journal of Technology Enhanced Learning, 15*(3), 266–285. https://doi.org/10.1504/IJTEL.2023.131865

Kastner-Hauler, O., Tengler, K., Sabitzer, B., & Lavicza, Z. (2022). Combined effects of block-based programming and physical computing on primary students' computational thinking skills. *Frontiers in Psychology, 13*, 875382. https://doi.org/10.3389/fpsyg.2022.875382

Katterfeldt, E.-S., Cukurova, M., Spikol, D., & Cuartielles, D. (2018). Physical computing with plug-and-play toolkits: Key recommendations for collaborative learning implementations. *International Journal of Child-Computer Interaction, 17*, 72–82. https://doi.org/10.1016/j.ijcci.2018.03.002

Kwon, K., Ottenbreit-Leftwich, A. T., Brush, T. A., Jeon, M., & Yan, G. (2021). Integration of problem-based learning in elementary computer science education: Effects on computational thinking and attitudes. *Educational Technology Research and Development, 69*, 2761–2787. https://doi.org/10.1007/s11423-021-10034-3

Love, T. S., & Asempapa, R. S. (2022). A screen-based or physical computing unit? Examining secondary students' attitudes toward coding. *International Journal of Child-Computer Interaction, 34*, C. https://doi.org/10.1016/j.ijcci.2022.100543

Min, S. H., & Kim, M. K. (2021). Developing children's computational thinking through physical computing lessons. *International Electronic Journal of Elementary Education, 13*(2), 183–198.

Neutens, T. (2022). *Physical computing in primary and secondary education: Techniques and tools for teaching and assessment*. Ghent University.

Nja, C. O., Orim, R. E., Neji, H. A., Ukwetang, J. O., Uwe, U. E., & Ideba, M. A. (2022). Students' attitude and academic achievement in a flipped classroom. *Heliyon, 8*(1), e08792. https://doi.org/10.1016/j.heliyon.2022.e08792

Ntourou, V., Kalogiannakis, M., & Psycharis, S. (2021). A study of the impact of Arduino and visual programming in self-efficacy, motivation, computational thinking and 5th grade students' perceptions on electricity. *EURASIA Journal of Mathematics, Science and Technology Education, 17*(5). https://doi.org/10.29333/ejmste/10842

Papavlasopoulou, S., Sharma, K., & Giannakos, M. N. (2018). How do you feel about learning to code? Investigating the effect of children's attitudes towards coding using eye-tracking. *International Journal of Child-Computer Interaction, 17*, 50–60. https://doi.org/10.1016/j.ijcci.2018.01.004

Petousi, V., & Sifaki, E. (2020). Contextualizing harm in the framework of research misconduct. Findings from discourse analysis of scientific publications. *International Journal of Sustainable Development, 23*(3/4), 149–174. https://doi.org/10.1504/IJSD.2020.10037655

Przybylla, M., & Romeike, R. (2018). Impact of physical computing on learner motivation. In *Proceedings of the 18th Koli calling international conference on computing education research* (pp. 1–10).

Sáez-López, J., Román-González, M., & Vázquez-Cano, E. (2016). Visual programming languages integrated across the curriculum in elementary school: A two year case study using "Scratch" in five schools. *Computers & Education, 97*, 129–141. https://doi.org/10.1016/j.compedu.2016.03.003

Salam, S. (2022). A systemic review of problem-based learning (PBL) and computational thinking (CT) in teaching and learning. *International Journal of Humanities and Innovation (IJHI), 5*(2), 46–52. https://doi.org/10.33750/ijhi.v5i2.145

Salta, K., & Koulougliotis, D. (2015). Assessing motivation to learn chemistry: Adaptation and validation of science motivation questionnaire II with Greek secondary school students. *Chemistry Education Research and Practice, 16*, 237–250.

Sentance, S., Waite, J., Hodges, S., MacLeod, E., & Yeomans, L. E. (2017). "Creating Cool Stuff"—Pupils' experience of the BBC micro:bit. In *Proceedings of the 48th ACM technical symposium on computer science education: SIGCSE 2017*.

Tai, R. H., Ryoo, J. H., & Skeeles-Worley, A., et al. (2022). (Re-)Designing a measure of student's attitudes toward science: a longitudinal psychometric approach. *IJ STEM Ed, 9*(12). https://doi.org/10.1186/s40594-022-00332-4

Theodoropoulos, A., Leon, P., Antoniou, A., & Lepouras, G. (2018). Computing in the physical world engages students: Impact on their attitudes and self-efficacy towards computer science through robotic activities. In *Proceedings of the 13th workshop in primary and secondary computing education*.

Vermillion, J., & de Salvatierra, A. (2019). Physical computing, prototyping, and participatory pedagogies; make-a-thon as interdisciplinary catalyst for bottom-up social change. In *Proceedings of the 2019 eCAADe + SIGraDi conference: Architecture in the age of the 4th industrial revolution* (pp. 359–366).

Vordou, E., & Romero, M. (2021). *Supporting grammar language through a Micro: Bit activity*. [Doctoral dissertation, UCA—INSPE Académie de Nice]. https://hal.archives-ouvertes.fr/hal-03408009v2

Weinburgh, M., & Steele, D. (2000). The modified attitudes toward science inventory: Developing an instrument to be used with fifth grade urban students. *Journal of Women and Minorities in Science and Engineering, 6*(1), 87–94. https://doi.org/10.1615/JWomenMinorScienEng.v6.i1.50

Chapter 12
Assessment of Computational Thinking Skills

Hüseyin Can Şenel and Selma Şenel

Abstract In this chapter, approaches used to assess computational thinking skills will be discussed. Good examples of these approaches also form an important aspect of the chapter. As a first approach, we will focus on scales based on the individual's perception of his/her computational thinking skill. These scales are frequently rating scales such as Likert type. Another approach is maximum performance tests to measure computational thinking. Achievement tests can be used with items that examine the individual's computational thinking skills. Various types of items are used in tests, such as multiple-choice items or open-ended items. It is considered that these approaches will not be sufficient to measure computational thinking, which is a high-level skill. As a result, and as a third approach, rubric scoring the product that emerges from a work such as a performance task or project in which the individual will reveal his/her computational skills. A variation on the performance-based approach is application-based assessment. The performance of the individual is evaluated on an application that enhances computational thinking skills, such as Scratch and Alice. Computational thinking skill will be assessed or reported by examining the source code of a project prepared in the application. Methods that blend four approaches in various ways are also observed as a fifth approach. In this chapter, these approaches will be introduced and examples in the literature will be presented. Common and eligible examples will also be discussed in the context of design-based learning. It can be concluded that computational thinking skills cannot be measured with a single assessment method. An assessment system that includes more than one method is recommended for deeper analyze of computational thinking skill. It is also evident that this system should include open-ended or performance-based approaches rather than multiple-choice items, as computational thinking is a high-level skill. Therefore, open-ended or performance-based approaches are emphasized in the chapter.

H. C. Şenel (✉)
Department of Computer Technology, National Defense University, Balıkesir, Türkiye
e-mail: huseyincan.senel@msu.edu.tr

S. Şenel
Department of Educational Sciences, Balıkesir University, Balıkesir, Türkiye
e-mail: selmasenel@balikesir.edu.tr

M. Saritepeci and H. Yildiz Durak (eds.), *Integrating Computational Thinking Through Design-Based Learning*, https://doi.org/10.1007/978-981-96-0853-9_12

Keywords Test for CT skills · CT assessment · Tests for perceived CT

Introduction

Over the past decade, computational thinking (CT) has received a great deal of attention from computer scientists, educators and researchers (Shute et al., 2017; Tsai et al., 2021). On the other hand, CT has not yet reached a scientific accumulation where there is agreement on its definition, theoretical dimensions or components (Kalelioglu et al., 2016). CT is focus of research for the last decade and getting more and more attention however there is no concensus about what CT contains, and there is a lack about the methods or strategies for assessing CT (Brennan & Resnick, 2012). This is, of course, due to the fact that CT is a structure that is formed in today's world, and it develops day by day.

Assessment of CT involves evaluating CT skills in educational contexts using various assessment constructs and types. The literature review conducted by Tang et al. (2020) proved that more CT assessments are needed for high school, college students, and teacher professional development programs.

Assessment of CT is in varied forms. While some researchers investigate aspects of computer science such as algorithms, programming, debugging, abstraction as CT assessment, some researchers focus on more comprehensive psychological constructs such as critical thinking, problem solving, digital literacy and STEM (D. Wang et al., 2022). On the other hand, confining computational thinking with computer science concepts may yield the result of limited representation of students' learning and participation for CT. Brennan and Resnick (2012) frames CT with three dimensions as computational concepts, computational practices and computational perspectives and offer to evaluate CT as an integrated concept. Computational concepts are the foundational terms typically applied by designers in the context of programming. Computational perspectives may be defined as an individual's personal interpretation of themselves and their surrounding environment in relation to their own program design. It is about how the designer perceives outer world. Computational practices are methodologies that are employed during the process of software development. These practices encompass a range of techniques, including testing, debugging, remixing, and the utilization of incremental and iterative approaches. Considering these dimensions measurement of CT must be complex and it must contain cognitive ability, the perception and the practice (Brennan & Resnick, 2012).

Studies on the assessment of CT offer different classifications by considering variables such as sub-dimensions of CT, age group and tools used. There are also CT assessment tools for different levels of education, such as primary school, middle school, K-12, high school, higher education and teachers. Depending on the targeted psychological construct and the targeted group, there are also differences in the assessment approaches of CT (Román-González et al., 2022a, 2022b). The basic measurement methods do not vary according to a specific concept, but when focusing

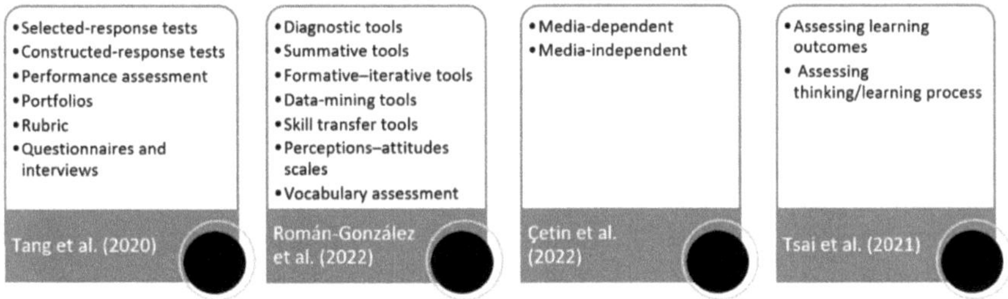

Fig. 12.1 Classifications of assessment of CT

on the assessment of CT, varied classifications stand out (Çetin et al., 2020; Román-González et al., 2022a, 2022b; Tang et al., 2020; Tsai et al., 2021). A schematized form of some of these classifications is presented in Fig. 12.1.

As presented in the figure, one of the latest classifications in the literature is presented by Román-González et al. (2022a, 2022b). They classified assessment of CT according to "evaluative approach" and grouped this approach into seven. Another noteworthy classification is based on "media-dependent and media-independent"; according to whether the measurement tool is required for the student having programming skills or whether students reply to tools' items without prior programming skills (Çetin et al., 2022). Tsai et al. (2021) focused on the measured characteristics in their classification. They defined the information about the individual's self-perception of computational thinking with tools such as surveys, self-reported scales or interviews related to CT competencies as "Assessing thinking/learning process". They referred to the measurement of sub-CT competencies such as syntax, data, and algorithms as "assessing learning outcomes". On the other hand, Tang et al. (2020) focused to the constructs to be assessed and classified commonly used tools considering in-class activities and sub-factors of CT.

In this study, the classification presented in Fig. 12.2 will be used. Differentiating from literature (Çetin et al., 2020; Román-González et al., 2022a, 2022b; Tang et al., 2020; Tsai et al., 2021) we reconsider both the theory and practice when we re-design and classify assessment of CT. We especially focused on the studies of Tang et al. (2020) and Brennan and Resnick (2012) and tried to band them together as the practical side and theoretical side respectively. Considering Brennan and Resnick (2012)'s research, we considered CT as an integrated concept and re-think as CT assessment must contain computational concepts, computational practices and computational perspectives. To be able to assess these dimensions of CT, Tang et al. (2020) offers a wide range of tools from selected-response tests to interviews and questionnaires. From another viewpoint to CT assessment, we notice that variables like attitude, perception or belief is some of the commonly researched variables related with CT (Román-González et al., 2022a, 2022b). Considering the dimensions of CT, the latest tools that current literature use for assessing CT and the research that "surrounding variables" like attitude, perception or belief we classify and examine the assessment of CT into four sub-headings as (a) assessment of

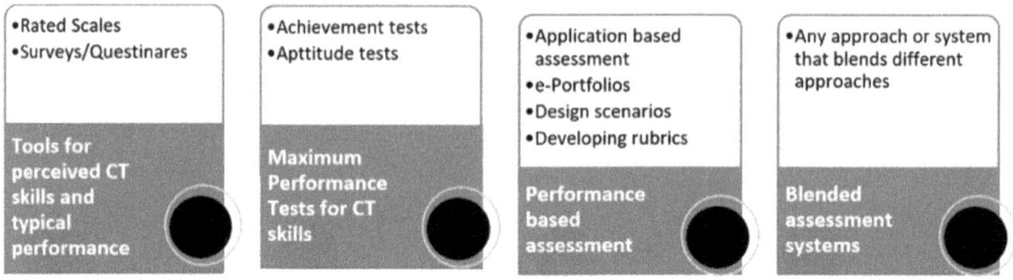

Fig. 12.2 Classification of assessment of CT in this study

perceived CT skills, (b) maximum performance tests for CT skills, (c) performance-based CT assessment and (d) blended assessment systems. Following headings will include examples of each category of CT assessment.

According to our classification, firstly, tools for perceived CT skills, which refers to the means by which the individual's perception of his/her own CT skills is determined through rated scales, will be discussed. In this part, addition to the tools to collect information about students' perceptions of CT skills, tools to assess typical performance such as CT interest, attitude and motivation also be discussed. Second, we will consider tests that aim to assess individuals' maximum performance in CT skills. These tests are achievement tests administered before or after a CT training program/intervention, or an aptitude test that aim to determine the individual's CT competencies. These instruments may include selected-response or constructed-response items. In the following section, we will discuss performance-based assessment tools that are aimed to measure higher level skills, which should be particularly emphasized in CT measurement. Performance-based assessment tools include techniques used to evaluate products in which an individual demonstrates CT skills. Examples of performance-based assessment can be application-based assessment tools such as Dr. Scratch where CT applications are automatically assessed, portfolios where different types of products are brought together, or design scenarios where students are given semi-structured scenarios. Considering design-based learning, this approach to assess CT is more prominent, focusing on specific products or designs of students. Finally, we must note that CT is a complex construct, and it is not possible to assess the CT with a single assessment approach and more comprehensive and hybrid systems have been produced in latest studies (Fang et al., 2022; Rottenhofer et al., 2022).

Tools for Perceived CT Skills and Typical Performance

Rating scales are often used to measure the perceived abilities of individuals. Respondents are presented with a statement about a particular skill and asked to indicate the extent to which they possess that skill. This means a perceived measurement based on the respondent's judgement. The results show the individual's perception of their competence, not their real competence or skill aboute the measured construct.

Attitudes of students towards CT or learning CT skills is one of mostly researched subjects. Hooshyar et al. (2021) focused on the learning attitude towards CT in an adaptive educational computer game context. Using a validated learning attitude questionnaire as pre-test and post-test, writers proved that use of adaptive games may enhance students' learning attitudes towards CT better than classical in-class approaches, especially for the ones with higher prior learning attitudes.

Surveys are another kind of instrument which are commonly used for non-cognitive variables such as motivation and attitudes towards CT. Researchers use surveys to explore the use of CT skills in various contexts. For example, study of Rottenhofer et al. (2022) focused on the use of CT skills in approaching linguistic learning tasks in the foreign language classroom and they used a survey to detect students' use of learning strategies linked to computational thinking. Another research used surveys and assessed the level of computational thinking skills among high school students, finding that the skill level was high and that female students had higher computational thinking skills than male students (Zakaria & Iksan, 2020).

Mostly used measurement tools for assessing perceived CT skills, their target group and scope are presented in Table 12.1. In selecting the measurement tools that will serve as exemplars for this section, two criteria were considered: the number of citations of the research in which the measurement tool was published and its focus on different age groups. It is of great importance to present validity and reliability evidence for these kind of assessment tools (Cutumisu et al., 2019). CT is a construct that is still developing and has managed to be an important subject of research. Determining whether CT is a critical variable for learning will direct the future of this phenomenal construct. Therefore, it is critical to use valid and reliable measurement tools in CT research (Jong et al., 2020). Table 12.1 presents some tools frequently used in the literature and some of their basic information. Some validity and reliability evidence, if available, is also included.

As Table 12.1 summarizes, literature offer measurement tools for assessing variables such as self-efficacy (perceived CT skills), interest, attitude and motivation. CT has gaining more interest over the last decade and having a wide range of instruments enables researchers to make creative designs to explain CT. Additionally, we can notice that these tools target participant from elementary schools to adult teachers. At this point, we only need validity and reliability proofs of these tools for being sure that accurate measurements are being made. While most of the measurement tools are presenting proof, some research has missing values for reliability and validity. Table 12.2. Presents sample items from mostly cited tools.

In Table 12.2, it can be observed that the items test the individual's own opinions regarding his competence in certain areas. Since the test result represents self-perception, it can also be affected by the individual's personality perception. On the one hand, in self-perception scales, social desirability may cause an individual's tendency to portray himself as better than he/she is. Therefore, the results of maximum performance tests are important for directly measuring CT skills.

Table 12.1 Tools for perceived CT skills and typical performance

Developers	Factors	Target group	Number of items	Evidence of validity/reliability
Kong et al. (2018)	Meaningfulness Impact Creative self-efficacy Prog. self-efficacy Interest in programming Attitude toward collaboration	Fourth/fifth and sixth graders	15 5-point Likert scale (1-Strongly disagree; 5-Strongly agree)	Confirmatory factor analysis, RMSEA = 0.058 No clue for reliability
Tsai et al. (2021)	Abstraction Decomposition Algorithmic thinking Evaluation Generalization	Eighth/ninth graders	19 5-point Likert scale (1-not agree at all)	Cronbach's alpha = 0.91 Exploratory factor analysis, total explained variance of 64.03%
Korkmaz et al. (2017)	Creativity Cooperativity Algorithmic thinking Critical thinking Problem solving	Higher education	29 5-point Likert scale (1-never; 5-always)	Exploratory factor analysis: total explained variance of 56.12% Confirmatory factor analysis, RMSEA = 0.062 Item discrimination analysis test-retest reliability = 0.512 Cronbach's alpha = 0.82
Gülbahar et al. (2019)	Algorithm design Problem solving Data processing Basic programming Self confidence	Fifth and sixth graders	36 3-point Likert scale (1-yes; 3-no)	Exploratory factor analysis: total explained variance of %53.102 Confirmatory factor analysis, RMSEA = 0.04 Item distinctiveness analyses Cronbach's alpha = 0.94

Table 12.2 Sample items for tools for perceived CT skills and typical performance

Developers	Factors	Sample items
Tsai et al. (2021)	Abstraction Decomposition Algorithmic thinking Evaluation Generalization	I usually try to find the key points of a problem I usually think of the structure of a problem I usually try to lay out the steps of a solution I tend to find a correct solution for a problem I tend to solve a new problem according to my experience
Korkmaz et al. (2017)	Creativity Algorithmic thinking Cooperativity Critical thinking Problem solving	I like the people who are sure of most of their decisions I can immediately establish the equity that will give the solution of a problem I like experiencing cooperative learning together with my group friends It is fun to try to solve the complex problems I cannot apply the solution ways I plan respectively and gradually
Kong et al. (2018)	Meaningfulness Impact Creative self-efficacy Programming self-efficacy Interest in programming Attitude toward collaboration in programming	Programming is useful to me I want to use programming to help solve problems in the world It is important to be creative when you are programming I think of myself as someone who can program I am curious about the content of programming People ask me for help with computers a lot

Maximum Performance Tests for CT Skills

In maximum performance tests, the competences of the respondant are tried to be estimated through the responses given to the items presented, not the respondant's perception of his/her competences. The aim of maximum performance tests is to directly determine the performance in the related field. When CT is considered, it is aimed to measure the competence of the individual in dimensions such as comprehension, analysis, evaluation and problem solving, which are accepted as competencies related to CT.

Maximum performance tests can be developed as an achievement test or as an aptitude test. If developed as an achievement test, it is intended to determine, after a CT enhancing programme, to what extent the objectives of the CT education program (intervention) have been achieved. The achievement test can also be used to narrow down the educational objectives prior to an education program/intervention or to determine which objectives have already been achieved. It can also be used to compare the levels of competence reached at the end of the education program/ intervention.

Aptitude tests are developed to reveal the general CT potential or strength of individuals. There is no educational content focus here. What is meant by the general CT potential of individuals is their ability to acquire knowledge and skills through education. However, these two maximum performance test types may have similar or overlapping scopes.

Tests for CT skills have been developed in various studies. The study of Hu and Wu (2023) developed a CT test that focuses on four key CT concepts: decomposition, data representation, algorithms, and pattern generalization. As stated above, researchers address varied constructs to assess CT. Algorithmic thinking is one of the common constructs considered related with CT. As a development and validation study, Lafuente Martínez et al. (2022) provided a reliable and valid tool to evaluate CT skills of adults. Another study by, Wang et al. (2023) proposed multidimensional test methods for CT assessment, combining question tests and programming tests with scale surveys. These methods allow for a comprehensive evaluation of learners' CT skills from both qualitative and quantitative perspectives. Overall, these studies highlight the importance of developing reliable and comprehensive assessments to evaluate CT skills in different educational contexts. Table 12.3 summarizes the characteristics of maximum performance tests used for assessment of CT. We again focused of citation rank of the research and their focus to age groups in selection.

Upon reviewing the table, several critical conclusions can be drawn. Some of these conclusions show promise for CT, which has yet to have its definition and components fully determined. However, others indicate that much work still needs to be done, particularly in measuring CT.

Firstly, to explain the positive results, it is possible to say that there is a consensus in the literature about how CT can measure the components related to computer science. All of the classical tests include similar computer science concepts such as algorithms, variables, loops, series and programming procedures such as modelling, sequencing and debugging. In addition, we notice that studies have been conducted for a wide age range from preschool age groups to university students. Moreover, researchers have tried to measure with different item types by using both open-ended and multiple-choice items. The deficiencies in the validity and reliability results of some of the analysed studies stand out as the biggest handicap for the measurement of CT. Concerns about validity and reliability were similarly expressed by Demir and Seferoğlu (2017).

Table 12.3 Maximum performance tests for CT skills

Developers	Factors/ components	Participants	Items	Validity and reliability
Relkin et al. (2020) -TechCheck	Algorithms Modularity Control structures Representation Hardware/ software Debugging	5–9 years old	15 multiple choice items	Cronbach's alpha = 0.68 Criterion validity Fleiss' Kappa = 0.63 Item information and item characteristic curves Criterion validity r = 0.53
Tran (2019)	Sequence Algorithms Looping Debugging Conditionals	Elementary students	10 open-ended items	Pilot testing No evidence
Cetin and Andrews-Larson (2016)	Sorting algorithms	College students	10 open-ended items	Content validity by experts Inter-rater reliability = 0.91
Román-González et al. (2017) CTt	Basic directions and sequences Loops Conditionals Functions	10–16 years old	28 multiple choice items	Cronbach's alfa = 0.79 Criterion validity

Performance Based Assessment

Performance based assessment is an assessment approach in which an individual is expected to show a performance (design, project or application) and this performance is scored. The developed design or program is scored through a rubric. If the design is developed in an environment/tool, this scoring can also be done automatically. It is also emphasized in the literature that smart systems that automatically measure these skills through the tools used by students should be produced (Wang et al., 2022).

In this chapter, performance-based assessment is discussed under four sub-headings: (1) application-based assessment, (2) e-portfolios, (3) design scenarios and (4) rubric development.

Application Based Assessment

Since the structure to be measured is CT, it seems more feasible to carry out application-based assessment. The scoring/grading is done according to the level of completion of different tasks within a given CT project tool or according to the developed projects or designs. As this is a system design, there is also an automatic evaluation for these systems. The evaluation dimensions and scales of the systems are also examined in detail. Examples of application-based assessment tools are reported in this section.

Dr. Scratch (https://www.drscratch.org/) (Moreno-León & Robles, 2015): Scratch is a web-based programming tool for youngers which let them to create their own projects (games, animations or entertaining scenarios), and then publicly share with peers (Resnick et al., 2009). It is a groundbreaking tool for K-12 coding education and very popular among youngsters. Among CT research, Scratch is frequently used. Dr. Scratch is a web-based assessment tool. It gives students a CT score and provides feedback on their projects once they upload or link to the programs they have developed (Moreno-León & Robles, 2015). Dr. Scratch can evaluate beginner or expert level projects. Dr. Scratch assess student projects in terms of abstraction, logical thinking, mapping, flow control, user interaction and data representation. Dr. Scratch also reviews students' coding mistakes. It also checks for bad programming habits such as keeping unused code or using unnecessary code blocks. Dr. Scratch evaluates each program according to following seven concepts: abstraction and problem decomposition, logical thinking, synchronization, parallelism, algorithmic notions of flow control, user interactivity and data representation. Developers gets 0–3 points for each of these concepts and assigned a CT score (max 21). Dr. Scratch produces a result report for each project. As CT score and proficiency increases, Dr. Scratch is able produce a feedback report with detailed information including CT skills and bad programming habits. Students can use this feedback to improve their programming and computational thinking skills.

Bebras (https://www.bebras.org/) (Dagiene & Futschek, 2008): The Bebras is an international organization focusing to enhance Informatics (Computer Science, or Computing) and computational thinking among students from 6 to 18 years old. The main aim of Bebras to initiate students' interest and development in informatics and computational thinking. It offers yearly nation-wide challenges, targeting to promote problem solving skills of students besides computational concepts. Bebras task items encourages students to break down complex tasks into simpler components using algorithms, patterns, generalisation and abstraction. A latest study used a selection of tasks of the 'The Bebras Computational Thinking Challenge' as CT measure to examine whether STEM learning attitude predicts CT skills of primary school students. The results indicated that attitudes towards learning in STEM were a significant predictor of CT skills (Sun et al., 2021).

Fairy Assessment (Werner et al., 2012): This is a performance assessment tool developed for measuring CT for K-12 students through a game programming class. This tool is build up on to assess comprehension, design and problem solving skills of middle school students referencing Chain of Cognitive Accomplishments (Linn et al., 1991). Using Alice, a Java-like programming environment, Fairy assessment is aimed to analyze the two components of CT as "thinking algorithmically" and "making effective use of abstraction and modeling". In 3 different tasks which has two fairies as avatars, Fairy Assessment targets to measure certain aspects of CT in each task.

Ninja Code Village (NCV) (Ota et al., 2016): Similar to Fairy Assessment, Ninja Code Village aims students to foster abstraction, modelling and decomposition competencies. As the Japan Government plan to integrate programming education in primary schools, NCV was one the preparations as a material. The assessment part of NCV is based on a function analyser. This analyser is able to detect and indicate which functions are used in a program or which ones can be used further. It may also route students to other corresponding samples. Besides, it makes an automatic assessment based on Brennan and Resnick's (2012) framework especially by "computational concepts". These criterian (computational concepts) are conditional statements, loops, procedure, common procedure, data, events, parallelism and user interface. By monitoring the Scratch blocks for 8 different criteria, automatic assessment detects and gives feedback to both the student and tutor which "computational concepts" are used and how they are used. NVC provides a good example of formative automatic assessment providing instant feedback and motivation.

e-Portfolios

One of the assessment design for CT is portfolio assessment or e-portfolios. Tang et al. (2020) reported that more than one-third of the studies assess students' CT skills using portfolio assessment. E-portfolio actually is a performance assessment based on collecting and assessing different types of products which are created through a semester or a given period of time. This type of assessment is formative, purpose-driven, timetabled and the aim is to support students' productive learning and to monitor achievement of learning objectives.

Portfolio assessment in CT contains several kinds of products such as projects (Scratch or other coding environments), prototypes, flowcharts, paper–pencil based practices, or direct observations in classroom activities. For example, Fields et al. (2021) designed a three-sectioned portfolio (Product, Process, and Reflection) for students in order to support students' communication skills about CT. Students need to fulfill text-based and visual-based requirements. Text-based requirements are documenting their projects' design, physical form, and coded behavior while visual-based requirements are photos, drawings and circuit diagrams. Both kind of material together create portfolio and researchers evaluate each students' portfolio using rubrics. Rubrics or grading rubrics are checklists to elaborate if the criterion is achieved evaluation criteria and quality definitions are met. By the way they have a clear scoring strategy (Brookhart, 2018). For example, Adler et al. (2023) developed a CT rubric which combines critical CT concepts considering Bloom's

Taxonomy. They used rubric for pre-service teachers who are taking STEM education and reported that the newly developed CT rubric presented chance them to assess students' CT competencies with a single method across a variety of STEM branches.

As detailed above, Scratch is focus of many CT research since it is an available tool for coding and project development. Collecting Scratch products, storing them and assessing is a kind of e-portfolio assessment. In his recent research, Chaou (2020) collected Scratch Jr projects (saved files), take-home assignments and observations as components of portfolio assessment and used learning observation document (a kind of rubric) to be able to monitor students' skills in CT concept, CT practice, and active learning. This research is a good example for how to bring together a portfolio with different kind of products and assess them with well-described rubrics.

As a limitation for portfolio assessment, students and evaluators needs time and energy (Lam, 2020). Portfolios may contain several type of products and these products are needed to be prepared by students and examined by the tutors in detailed since any of them may give clues about the development of the student. Portfolio assessment is a process rather than a one-time-assessment. Second, to conduct portfolio assessment a grading rubric and specific evaluation standards should be established and developed correctly and appropriately (Yoo et al., 2020).

We must remember that a collection of product presents chance to see how perceiving and learning increases over time and makes evaluation even detailed (Brennan & Resnick, 2012). Second, portfolio assessment provides teacher feedback and critique which are highly valuable for developing new viewpoints and increasing motivation.

Design Scenarios

Design scenarios is one of the assessment approaches that (Brennan & Resnick, 2012) offered to assess CT. In this approach, students are given programming design scenarios developed by researchers for the classroom environment. These scenarios are created at different levels of difficulty. Students choose one scenario from each level. Students are asked questions about what the project does (1), how it can be improved (2), debugged and fixed (3) and add new features (4) (Brennan & Resnick, 2012). By answering these four types of questions students are encouraged to make critiquing, extending, debugging, and remixing.

In a design scenario assessment research (Zhong et al., 2016), six different semifinished programming projects, reflection and design reports are used. Students are presented Alice-based semi-finished projects and encouraged to complete projects. The aim is to enrich and assess students' ability of planning, sequencing, debugging, optimizing, etc. and use of computer programming components like variables and loops. Like (Brennan & Resnick, 2012) researchers aroused students by routing them to analyze semi-finished projects for debugging, fixing or troubleshooting. For example, they used reverse tasks (troubleshooting tasks) for CT assessment and open and semi-open tasks for CT assessment and found that ill-structured projects assess more dimensions of CT. Additionally, reflection and creative design reports were practicable to use in-class and gave feedback to teachers about learning diagnosis and how to guide teacher to find out students' thinking barriers.

Rubric

Rubrics are used for constructed-response items and performance-based tasks where manual scoring is required (Meier et al., 2006). Rubrics are created to score a project, design, or a long answer. Rubrics are crucial for ensuring scoring reliability and enabling developers to provide feedback (Li, 2022). Rubrics must be developed to enable different raters to score a product as accurately as possible.

A rubric can be either holistic or analytic: a holistic rubric assesses the whole process or product without giving specific feedback to students, whereas an analytic rubric first assesses separate, individual parts of the product or performance and then adds up the individual scores to give an overall score (Panadero & Jonsson, 2020). A holistic rubric is preferable when the project or task being measured does not have to be assessed separately through the determination of specific grades. However, if the characteristic to be measured has several dimensions, the analytic rubric should be preferred. Furthermore, as another differentiation, rubrics can be designed for singular or multiple tasks/assignments (Jonsson et al., 2007). In this case, rubric is named as task-specific rubrics or generic rubrics.

Regarding CT, it is unlikely that we can discuss a single task or dimension, as it is a complex and integrated concept. The tasks, assignments or projects as products of CT skills are not one-sided. They are multi-faceted in nature. In Table 12.4 we offer an example of a rubric for the evaluation of a beginner programmer CT assignment.

Above, we offer a sample rubric that addresses the CT indicators expected from the tasks or assignments. So far, we have reported that there have been a number of different approaches to the measurement of CT. At this point, it will be important to take into account how the researcher or teacher defines CT and which dimensions they address when determining the criteria to be addressed in the rubric. This example rubric has been developed on the basis of the Tsai et al. (2021) CT dimensions. For each dimension, the quality of the student product is rated on a three-point scale. A score of "0" is given if the student product does not contain any indicators related to the relevant dimension. The range of the scale can be changed by increasing or decreasing the scores. However, the provision of clear and objective explanations for each score in each dimension enhances the objectivity of the rubric in the scoring. In performance-based assessment and in tests with open-ended items, the scoring and the development of the rubric are as important as the task or the item.

Table 12.4 An analytic-generic rubric example for grading a CT assignment

Criteria	Proficiency level			Point
	Basic (1)	Developing (2)	Proficient (3)	
Abstraction	Focused on the details of the problem	Approached at some points holistically, generally focussed on the details	Analysed the problem from a holistic point of view	
Decomposition	Failed to recognise minor problems beneath the main problem	Recognised some of the minor problems beneath the main problem	Recognised all of the minor problems beneath the main problem	
Algorithmic thinking	The steps determined for the solution of the problem are not sufficient	Determined the basic steps for the solution of the problem	Established steps and relations to solve the problem effectively and efficiently (with minimum steps)	
Evaluation	A solution with low efficiency and performance was selected among similar options	A relatively effective solution has been selected. Not aware of more effective/ well-performing options	The most efficient and well-performing solution has been selected among various solutions	
Generalization	The possibility of transferring the developed solution to different problems is quite low	The developed solution is likely to be applied to different problems	The developed solution has a high probability of being utilised for a large number of problems	
			Total point	

Blended Assessment Systems

It is not possible to claim that every measurement method or approach is suitable for measuring every cognitive level. When considering Bloom's taxonomy as a basic framework, evaluation and synthesis can be highlighted as high-level skills. In the Haladayna taxonomy, critical thinking and problem-solving are considered skills of the same level. Particularly for measuring problem-solving or synthesis levels, items that require selection may not be appropriate. The use of various tools and methods to assess an individual's competence or perception of competence can provide a more accurate picture. Studies in the literature have also highlighted the importance of this necessity (Brennan & Resnick, 2012; Román-González et al., 2022a, 2022b).

At this point, the need for blended assessment systems to measure high-level skills such as CT emerges. Blended assessment systems in computational thinking involve the integration of multiple forms of assessment to provide a comprehensive assessment of student learning. This approach recognises the need to go beyond traditional assessment methods and incorporate different types of assessment to effectively assess computational thinking skills (Grover & Pea, 2015). The aim is to create a system that allows teachers to supervise the learning process and assess students individually, while providing an intuitive interface (Triswidrananta et al., 2022). The use of multiple measures or "systems of assessments" is recommended to assess the development of computational thinking skills, including both cognitive and non-cognitive aspects.

As one of the most cited studies, Grover and Pea's (2015) study may provide ideas for future research. Using multiple-choice quizzes, directed and open-ended programming tasks, and assessments, they aimed to assess cognitive skills such as algorithmic thinking skills and self-reported variables such as perceptions of computing. They used a total of seven different instruments: two surveys, two achievement tests and three types of achievement tests. Similarly, Fang et al. (2022) used a peer-assessed approach to improve students' computer programming skills and computational thinking awareness. Using a quasi-experimental design, they intervened in a programming course and used different instruments to assess CT-related variables such as achievement, computational thinking awareness, motivation and self-efficacy. In a blended system, they used an online peer assessment tool, an achievement test and a perceived CT scale.

However, there are challenges to develop such systems and applications that need to be addressed (Fang et al., 2022). First, the design and research of blended systems is quite complex. It requires the combined use of multiple types of methods and assessment tools, which involves both a difficult selection process to find appropriate methods and tools, and the time and effort required to implement them. A further challenge is the difficulty that students have in taking part in all of these activities or assessment processes. In conjunction with these reasons, blended systems require a long period of time to be implemented and for the results to be analysed.

Assessing Computational Thinking on Context of Design-Based Learning

Design-based learning (DBL) is a pedagogical approach that anchors learning in design activities, often in groups. It encompasses reflection-in-action, where learners reflect on their design experiences to enhance learning. DBL empowers teachers as intellectual designers, allowing them to plan lessons, implement instructional practices, and reflect on their practice for professional growth (Zhang et al., 2023). Design-based learning can be used to teach computational thinking by integrating interdisciplinary activities and problem-solving techniques. This approach involves using design-based learning models, such as the scientific cycle model, and incorporating STEAM education principles (Wang et al., 2022).

CT offers to produce assignments, projects, tasks, or task series. Producing practical solutions to real problems is a requirement for enhancing CT skills. This includes generating algorithms for scenario-based environments or design-based learning. Developing web-based projects using tools such as Alice or Scratch, gathering portfolios that include self-reports, assignments, or small-scale projects are steps that bring us into the complex world of DBL. DBL offers more opportunities than any other CT-enhancing activity. It provides the advantage of learning-by-doing and collaboration and represents freedom for both learners and instructors. Having more freedom in activities without clear designs or instructions can lead to chaos and a lack of structure. DBL employs complex learning designs and unconventional learning activities, and most learners and instructors may not be ready for such activities. Additionally, assessing these unfamiliar activities requires detailed study and observation. The use of multiple tools in combination can aid in the assessment of validity and reliability. Now we will summarize some of the DBL research for CT and their assessment approaches.

Jun et al. (2017) conducted a study to investigate the impact of DBL on the CT skills of primary school pupils. The study compared the effectiveness of DBL and traditional direct method group in improving CT skills. Both groups took a programming course using Scratch. The results showed that the DBL group outperformed the traditional group, indicating that DBL is a more effective method for improving CT skills. They used different type of tools such as self-efficacy and self-interest questionnaire and a self-CT test.

In a similar research, Saritepeci (2020) used measurement tools such as "Personal Information Form", "Computational Thinking Skill Scale (CTSS)" and "Reflective Thinking Skill Scale toward Problem Solving (RTSSPS)". As a good example of use of classical measurement methods, valid and reliable tools are used to compare two groups of high-school students. Completing fourteen weeks of traditional and DBL activities (designing of story boards on papers based on scenario, forming digital storytelling and QR codes, preparing infographics, etc.) results demonstrated that DBL has a significant positive development at CT levels. Besides DBL has a larger effect size on CT development compared to traditional instruction of computing.

A recent study examined the impact of DBL on the computational thinking skills of elementary students in a visual programming maker course using mixed-method research. The study found that students who engaged in DBL activities (hands-on projects and problem-solving tasks in this research) demonstrated enhanced skills in problem analysis, critical thinking, and solution development using computational thinking concepts. Additionally, the research showed that DBL facilitated the development of students' creativity, collaboration skills, and ability to apply CT in real-world contexts. Researchers used two different assessment tools as computational thinking test and artifacts (final product) (Matere et al., 2023).

Lyon and Magana (2021) researched how model-building activities may enhance CT skills of engineering students. This research is very critical in two ways. One is with its participants and second, the tools used for assessment of CT. In a capstone course, researchers used students' artifacts (an individual take home assignment) and an ill-structured programming file. Results proved that DBL design of the study elicited CT outcomes offered clues about the understanding of how CT takes place in the context of engineering education.

In another research subjecting DBL and critical thinking/problem solving abilities, Shanta and Wells (2022) focused on high school technology and engineering students. This research is a good example of integration of DBL learning and use of rubrics. After determining five key student abilities for assessment of problem-solving skills, researchers gave students well designed scenarios and tasks within the context of authentic engineering design. Then they developed a rubric to be able to make a reliable and valid measurement. Results indicated that students who take STEM content through design-based learning showed significantly better critical thinking/problem solving abilities in designing an engineering solution compared to traditional classroom instruction.

Above we summarize five different studies conducted by DBL. It is worth noting that each of these studies used different CT assessment tools. The first study employed perceived CT skills and typical performance tests, which constitute the initial part of our CT assessment classification. The second study utilised Maximum Performance Tests, which form the second part of our classification. Third, employed performance-based assessment tools, which constitute the third part of our classification. Fourth and fifth similarly used take-home assignments and scenarios while the latter is a good example of using rubrics.

As we mentioned in the title of blended assessment, there is no single most accurate instrument for measuring a construct. With the use of technology tools in education, course activities and research designs have diversified, and unique environments are emerging. Furthermore, CT is a concept that originates from technology. To ensure valid and reliable measurement of CT, it is crucial to accurately identify the sub-factors of CT and select appropriate measurement tools based on the research context. It is evident that employing multiple measurement tools to assess various constructs, particularly in CT studies that involve blended and DBL, can enhance the study's outcomes.

Conclusion

CT is still not well defined (Korkmaz et al., 2017). This lack of definition hinders the full development of CT (Román-González et al., 2022a, 2022b). CT is a complex term and should not only revolve around concepts that are insufficiently represented in other aspects of design, learning and participation (Brennan & Resnick, 2012.

Although research on CT and assessment of CT is extremely popular, CT is not yet matured as it is a newborn concept. In this chapter, we have focused on the assessment of CT and tried to bring together the theory with current practices. The tools used to measure CT are widespread. Different types of assessment tools are used to assess CT, ranging from self-perceived measures to computer science achievement tests, from rubrics to automated assessment tools. Formative and summative measurement tools are also used. This diversity and complexity arises from the fact that the concept of CT is multidimensional and that measurement tools evolve with technology.

Summary of Key Points

We have developed a classification for measuring CT by combining theory and practice and considering its emergence with technological development. The first section of our classification covers tools for perceived CT skills and typical performance, which are used to gather participants' opinions and perceptions of themselves.

Second, we clarified the maximum performance tests for CT skills. Computational concepts appear as an important component of CT and these tests are mostly computer science achievement tests. Algorithms, variables, conditional expressions, loops, which are the basis of computer science, form the scope of these tests. These tests are given to the participants as a pre-test and/or post-test and the effects of the interventions are studied.

Thirdly, we mentioned CT measures that use performance-based assessment. CT is a term that originated in computer science and is often associated with programming education. Projects/software that encourage programming, such as Scratch and Alice, e-portfolios consisting of various media that can be easily stored and transported with the help of technology, scenario and game-based environments that encourage students' creativity are used in performance-based measurement.

Finally, by reviewing the studies on the measurement of CT in DBL settings, we emphasised the importance of examining CT in the context of research and emphasized the importance of determining and defining the measured construct and the need of using different types of measurement tools together.

Implications for Future Practice and Research

The term CT, which has been identified as a key concept for addressing everyday challenges, has the potential to facilitate a transformative shift in the way we educate at the school level. It has the capacity to engender a novel mode of thinking for all of us. As previously stated, there are a number of ways in which CT can be assessed. Nevertheless, the measurement of CT occurs subsequent to the act of teaching, and there is a willingness to measure what we teach. In this section, we present recommendations for future measurement.

Firstly, it is important to recognize that the measurement of computational thinking (CT) is a complex process. There is no straightforward, definitive approach. Secondly, the measurement of CT is of paramount importance for the advancement of technology, the resolution of complex issues and the continuous growth and improvement of individuals. In order to achieve this, it is necessary to obtain support for more far-reaching decisions, such as those pertaining to nationwide practices. Thirdly, there is a necessity for the development of highly proficient educators who are capable of providing effective instruction in CT. It is only through the development of the teacher that students can be assisted with complex activities, which may enable them to work in groups and to think in a manner that is analogous to that of a computer when they are faced with different kinds of problems. Finally, well-designed research is required which reports on each of the activities employed for the enhancement of CT, together with a detailed description of the measurement steps.

As a last and critical research subject, we point artificial intelligence (AI). The application of AI is accelerating in measurement studies, as it is in all fields of research. The utilization of AI in item generation or scoring of responses is a topic that has been extensively investigated. The utilization of AI in the assessment of CT within the context of design-based learning has the potential to inform future practices and applications. As an extension of performance-based assessments and blended assessment approaches, it may be posited that individuals are not presented with the same achievement test on each occasion during the measurement process, but rather with items generated by AI. Furthermore, the application of adaptive tests, which are tailored to the progress of the student, may also be employed to assess an individual's ability with greater accuracy and reliability.

The enhancement of CT and the measurement of CT are not easy to follow. It is imperative that we develop CT activities and measurement tools that extend beyond the conventional in-class activities and self-response tests. It would be beneficial to allow teachers additional time to design more complex activities. It is also important to consider the specific aim, time constraints and organizational requirements.

References

Adler, R. F., Hibdon, J., Kim, H., Mayle, S., Pines, B., & Srinivas, S. (2023). Assessing computational thinking across a STEM curriculum for pre-service teachers. *Education and Information Technologies, 28*(7), 8051–8073. https://doi.org/10.1007/s10639-022-11508-4

Brennan, K., & Resnick, M. (2012). New frameworks for studying and assessing the development of computational thinking. *AERA*, 1.

Brookhart, S. M. (2018). Appropriate criteria: Key to effective rubrics. *Frontiers in Education, 3*(April). https://doi.org/10.3389/feduc.2018.00022

Cetin, I., & Andrews-Larson, C. (2016). Learning sorting algorithms through visualization construction. *Computer Science Education, 26*(1), 27–43. https://doi.org/10.1080/08993408.2016.1160664

Çetin, İ., Otu, T., & Oktaç, A. (2020). Adaption of the computational thinking test into Turkish. *Turkish Journal of Computer and Mathematics Education, 11*(2), 343–360. https://doi.org/10.16949/turkbilmat.643709

Çetin, İ., Şendurur, P., & Otu, T. (2022). Tech Check İsimli Bilgi İşlemsel Düşünme Testlerinin Türkçeye Uyarlanması. *Journal of Instructional Technologies and Teacher Education, 11*(2), 16–27. https://doi.org/10.51960/JITTE.1102904

Cutumisu, M., Adams, C., & Lu, C. (2019). A scoping review of empirical research on recent computational thinking assessments. *Journal of Science Education and Technology, 28*(6), 651–676. https://doi.org/10.1007/s10956-019-09799-3

Dagiene, V., & Futschek, G. (2008). Bebras international contest on informatics and computer literacy: Criteria for good tasks. *Lecture Notes in Computer Science (Including Subseries Lecture Notes in Artificial Intelligence and Lecture Notes in Bioinformatics), 5090 LNCS*(February), 19–30. https://doi.org/10.1007/978-3-540-69924-8_2

Demir, Ö., & Seferoğlu, S. S. (2017). Yeni Kavramlar, Farklı Kullanımlar: Bilgi-İşlemsel Düşünmeyle İlgili Bir Değerlendirme (Novel concepts, different usages: An evaluation on computational thinking). *Eğitim Teknolojisi Okumaları 2017 (Readings of Educational Technologies 2017)*, 801–830. https://www.researchgate.net/publication/318908996_Yeni_Kavramlar_Farkli_Kullanimlar_Bilgi-Islemsel_Dusunmeyle_Ilgili_Bir_Degerlendirme

Fang, J., Shao, D., & Hwang, G. (2022). *From critique to computational thinking: A peer-assessment-supported problem identification, flow definition, coding, and testing approach for computer programming instruction. 60*(5), 1301–1324. https://doi.org/10.1177/07356331211060470

Fields, D., Lui, D., Kafai, Y., Jayathirtha, G., Walker, J., & Shaw, M. (2021). Communicating about computational thinking: Understanding affordances of portfolios for assessing high school students' computational thinking and participation practices. *Computer Science Education, 31*(2), 224–258. https://doi.org/10.1080/08993408.2020.1866933

Gülbahar, Y., Kert, S. B., & Kalelioglu, F. (2019). Bilgi Işlemsel Düşünme Becerisine Yönelik Oz Yeterlik Algisi Olcegi: Gecerlik ve Güvenirlik Calismasi. *Turkish Journal of Computer and Mathematics Education, 10*(1), 1–29. https://doi.org/10.16949/turkbilmat.385097

Hooshyar, D., Malva, L., Yang, Y., Pedaste, M., Wang, M., & Lim, H. (2021). An adaptive educational computer game: Effects on students' knowledge and learning attitude in computational thinking. *Computers in Human Behavior, 114*(September 2020), 106575. https://doi.org/10.1016/j.chb.2020.106575

Hu, C.-F., & Wu, C.-C. (2023). A computational thinking test for senior high school students. In *Proceedings of the 2023 conference on innovation and technology in computer science education* (Vol. 2, pp. 587–588). https://doi.org/10.1145/3587103.3594151

Jong, M. S. Y., Geng, J., Chai, C. S., & Lin, P. Y. (2020). Development and predictive validity of the computational thinking disposition questionnaire. *Sustainability (Switzerland), 12*(11). https://doi.org/10.3390/su12114459

Jonsson, A., & Svingby, G. (2007). The use of scoring rubrics: Reliability, validity and educational consequences. *Educational Research Review, 2*(2), 130–144.

Kalelioglu, F., Gulbahar, Y., & Kukul, V. (2016). *A framework for computational thinking based on a systematic research a framework for computational thinking based on a systematic research review. May.*

Kong, S. C., Chiu, M. M., & Lai, M. (2018). A study of primary school students' interest, collaboration attitude, and programming empowerment in computational thinking education. *Computers and Education, 127*, 178–189. https://doi.org/10.1016/j.compedu.2018.08.026

Korkmaz, Ö., Çakir, R., & Özden, M. Y. (2017). A validity and reliability study of the computational thinking scales (CTS). *Computers in Human Behavior, 72*, 558–569. https://doi.org/10.1016/j.chb.2017.01.005

Lafuente Martínez, M., Lévêque, O., Benítez, I., Hardebolle, C., & Zufferey, J. D. (2022). Assessing computational thinking: Development and validation of the algorithmic thinking test for adults. *Journal of Educational Computing Research, 60*(6), 1436–1463. https://doi.org/10.1177/07356331211057819

Lam, R. (2020). Writing portfolio assessment in practice: Individual, institutional, and systemic issues. *Pedagogies, 15*(3), 169–182. https://doi.org/10.1080/1554480X.2019.1696197

Li, W. (2022). Scoring rubric reliability and internal validity in Rater-mediated EFL writing assessment: Insights from many-facet Rasch measurement. *Reading and Writing, 35*(10), 2409–2431.

Linn, R. L., Baker, E. L., & Dunbar, S. B. (1991). Complex, performance-based assessment: Expectations and validation criteria. *Educational Researcher, 20*(8), 15–21.

Lyon, J. A., & Magana, A. J. (2021). The use of engineering model-building activities to elicit computational thinking: A design-based research study. *Journal of Engineering Education, 110*(1), 184–206. https://doi.org/10.1002/jee.20372

Matere, I. M., Weng, C., Astatke, M., Hsia, C.-H., & Fan, C.-G. (2023). Effect of design-based learning on elementary students computational thinking skills in visual programming maker course. *Interactive Learning Environments, 31*(6), 3633–3646. https://doi.org/10.1080/10494820.2021.1938612

Meier, S. L., Rich, B. S., & Cady, J. (2006). Teachers' use of rubrics to score non-traditional tasks: Factors related to discrepancies in scoring. *Assessment in Education, 13*(01), 69–95.

Moreno-León, J., Robles, G., & Román-González, M. (2015). Dr. Scratch: automatic analysis of scratch projects to assess and foster computational thinking. *RED. Revista de Educación a Distancia, 15*(46), 1–23. http://www.um.es/ead/red/46/moreno_robles.pdf

Ota, G., Morimoto, Y., & Kato, H. (2016). Ninja code village for scratch: Function samples/function analyser and automatic assessment of computational thinking concepts. In *Proceedings of IEEE symposium on visual languages and human-centric computing, VL/HCC, 2016-Nov*, pp. 238–239. https://doi.org/10.1109/VLHCC.2016.7739695

Panadero, E., & Jonsson, A. (2020). A critical review of the arguments against the use of rubrics. *Educational Research Review, 30*, 100329.

Relkin, E., de Ruiter, L., & Bers, M. U. (2020). TechCheck: Development and validation of an unplugged assessment of computational thinking in early childhood education. *Journal of Science Education and Technology, 29*(4), 482–498. https://doi.org/10.1007/s10956-020-09831-x

Román-González, M., Moreno-León, J., & Robles, G. (2022a). Combining assessment tools for a comprehensive evaluation of computational thinking interventions. In S.-C. Kong, & H. Abelson (Eds.), *Computational thinking education in K–12* (pp. 1–12). https://doi.org/10.7551/mitpress/13375.003.0002

Román-González, M., Moreno-León, J., & Robles, G. (2022b). Combining assessment tools for a comprehensive evaluation of computational thinking interventions. In S.-C. Kong, & H. Abelson (Eds.), *Computational thinking education in K–12* (pp. 79–98). https://doi.org/10.7551/mitpress/13375.003.0002

Román-González, M., Pérez-González, J. C., & Jiménez-Fernández, C. (2017). Which cognitive abilities underlie computational thinking? Criterion validity of the computational thinking test. *Computers in Human Behavior, 72*, 678–691. https://doi.org/10.1016/j.chb.2016.08.047

Rottenhofer, M., Kuka, L., Leitner, S., & Sabitzer, B. (2022). Using computational thinking to facilitate language learning: A survey of students' strategy use in Austrian secondary schools. *IAFOR Journal of Education, 10*(2), 51–70. https://doi.org/10.22492/ije.10.2.03

Saritepeci, M. (2020). Developing computational thinking skills of high school students: Design-based learning activities and programming tasks. *The Asia-Pacific Education Researcher, 29*(1), 35–54.

Shanta, S., & Wells, J. G. (2022). T/E design based learning: Assessing student critical thinking and problem solving abilities. *International Journal of Technology and Design Education, 32*(1), 267–285.

Shute, V. J., Sun, C., & Asbell-clarke, J. (2017). *Version of record*. https://www.sciencedirect.com/science/article/pii/S1747938X17300350

Sun, L., Hu, L., Yang, W., Zhou, D., & Wang, X. (2021). STEM learning attitude predicts computational thinking skills among primary school students. *Journal of Computer Assisted Learning, 37*(2), 346–358. https://doi.org/10.1111/jcal.12493

Tang, X., Yin, Y., Lin, Q., Hadad, R., & Zhai, X. (2020). Assessing computational thinking: A systematic review of empirical studies. *Computers & Education, 148*, 103798. https://doi.org/10.1016/J.COMPEDU.2019.103798

Tran, Y. (2019). Computational thinking equity in elementary classrooms: What third-grade students know and can do. *Journal of Educational Computing Research, 57*(1), 3–31. https://doi.org/10.1177/0735633117743918

Tsai, M.-J., Liang, J.-C., Hsu, C.-Y., & Tsai, J. (2021). The computational thinking scale for computer literacy education. *Journal of Educational Computing Research, 59*(4), 579–602. https://doi.org/10.1177/0735633120972356

Wang, D., Luo, L., Luo, J., Lin, S., & Ren, G. (2022). Developing computational thinking: Design-based learning and interdisciplinary activity design. *Applied Sciences (Switzerland), 12*(21). https://doi.org/10.3390/app122111033

Wang, J., Zhang, W., Zeng, X., & Li, P. (2023). A computational thinking assessment tool on text-based programming. In *2023 IEEE 12th international conference on educational and information technology (ICEIT)* (pp. 326–331). https://doi.org/10.1109/ICEIT57125.2023.10107885

Werner, L. W., Denner, J., & Hizuru Kawamoto, D. C. (2012). *The fairy performance assessment: Measuring computational thinking in middle school.*

Yoo, D. M., Cho, A. R., & Kim, S. (2020). Development and validation of a portfolio assessment system for medical schools in Korea. *Journal of Educational Evaluation for Health Professions, 17*, 1–7. https://doi.org/10.3352/JEEHP.2020.17.39

Zakaria, N. I., & Iksan, Z. H. (2020). Computational thinking among high school students. *Universal Journal of Educational Research, 8*(11 A), 9–16. https://doi.org/10.13189/ujer.2020.082102

Zhang, Z., Bekker, T., Markopoulos, P., & Skovbjerg, H. M. (2023). Supporting and understanding students' collaborative reflection-in-action during design-based learning. *International Journal of Technology and Design Education, 34*(1). https://doi.org/10.1007/s10798-023-09814-0

Zhong, B., Wang, Q., Chen, J., & Li, Y. (2016). An exploration of three-dimensional integrated assessment for computational thinking. *Journal of Educational Computing Research, 53*(4), 562–590. https://doi.org/10.1177/0735633115608444

MIX
Papier aus verantwortungsvollen Quellen
Paper from responsible sources
FSC® C105338

If you have any concerns about our products, you can contact us on
ProductSafety@springernature.com

In case Publisher is established outside the EU, the EU authorized representative is:
Springer Nature Customer Service Center GmbH
Europaplatz 3, 69115 Heidelberg, Germany

Printed by Libri Plureos GmbH
in Hamburg, Germany